98 Springer Series in Solid-State Sciences
Edited by Peter Fulde

Springer Series in Solid-State Sciences

Editors: M. Cardona P. Fulde K. von Klitzing H.-J. Queisser

Managing Editor: H. K.V. Lotsch Volumes 1–89 are listed at the end of the book

90 **Earlier and Recent Aspects of Superconductivity** Editors: J. G. Bednorz and K. A. Müller

91 **Electronic Properties of Conjugated Polymers III** Basic Models and Applications
Editors: H. Kuzmany, M. Mehring, and S. Roth

92 **Physics and Engineering Applications of Magnetism** Editors: Y. Ishikawa and N. Miura

93 **Quasicrystals** Editors: T. Fujiwara and T. Ogawa

94 **Electronic Conduclion in Oxides** By N. Tsuda, K. Nasu, A. Yanase, and K. Siratori

95 **Electronic Materials** A New Era in Materials Science
Editors: J. R. Chelikowsky and A. Franciosi

96 **Electron Liquids** By A. Isihara

97 **Localization and Confinement of Electrons in Semiconductors**
Editors: F. Kuchar, H. Heinrich, and G. Bauer

98 **Magnetism and the Electronic Structure of Crystals**
By V. A. Gubanov, A. I. Liechtenstein, and A.V. Postnikov

99 **Electronic Properties of High-T_c Superconductors and Related Compounds**
Editors: H. Kuzmany, M. Mehring, and J. Fink

100 **Electron Correlations in Molecules and Solids** By P. Fulde

101 **High Magnetic Fields in Semiconductor Physics III**
Quantum Hall Effect, Transport and Optics By G. Landwehr

102 **Conjugated Conducting Polymers** Editor: H. Kiess

103 **Molecular Dynamics Simulations** Editor: F. Yonezawa

104 **Products of Random Matrices in Statistical Physics**
By A. Crisanti, G. Paladin, and A. Vulpiani

105 **Self-Trapped Excitons** By K. S. Song and R. T. Williams

106 **Physics of High-Temperature Superconductors** Editors: S. Maekawa and M. Sato

107 **Electronic Properties of Polymers** Orientation and Dimensionality of Conjugated Systems
Editors: H. Kuzmany, M. Mehring, and S. Roth

108 **Site Symmetry in Crystals** Theory and Applications
By R. A. Evarestov and V.V. Smirnov

V. A. Gubanov · A. I. Liechtenstein · A. V. Postnikov

Magnetism and the Electronic Structure of Crystals

With 79 Figures

Springer-Verlag
Berlin Heidelberg New York
London Paris Tokyo
Hong Kong Barcelona Budapest

Professor Vladimir A. Gubanov
Dr. Alexandr I. Liechtenstein

Institute of Solid State Chemistry,
Russian Academy of Sciences,
Pervomajskaja 91, Jekaterinburg 620219, Russia

Dr. Andrei V. Postnikov

Institute of Metal Physics,
Russian Academy of Sciences,
S. Kowalewskoj 18, Jekaterinburg 620219, Russia

Series Editors:

Professor Dr., Dres. h. c. Manuel Cardona
Professor Dr., Dr. h. c. Peter Fulde
Professor Dr., Dr. h. c. Klaus von Klitzing
Professor Dr. Hans-Joachim Queisser

Max-Planck-Institut für Festkörperforschung, Heisenbergstraße 1
D-7000 Stuttgart 80, Fed. Rep. of Germany

Managing Editor:

Dr. Helmut K. V. Lotsch

Springer-Verlag, Tiergartenstraße 17
D-6900 Heidelberg, Fed. Rep. of Germany

ISBN-13: 978-3-642-84413-3 e-ISBN-13: 978-3-642-84411-9
DOI: 10.1007/978-3-642-84411-9

Library of Congress Cataloging-in-Publilcation Data
Gubanov, V. A. (Vladimir Aleksandrovich)
Magnetism and the electronic structure of crystals / V. A. Gubanov, A. I. Liechtenstein, A. V. Postnikov.
p. cm. -- (Springer series in solid-state sciences ; 98)
Includes bibliographical references and index.
ISBN-13: 978-3-642-84413-3
1. Energy-band theory of solids. 2. Solids--Magnetic properties.
3. Crystals--Magnetic properties. I. Likhtenshteĭn, A.
I. (Aleksandr Iosifovich) II. Postnikov, A.V. (Andreĭ Viktorovich) III. Title. IV. Series.
QC176.8.E4G83 1992 530.4'12--dc20 92-19761

Softcover reprint of the hardcover 1st edition 1992

Typesetting: Macmillan India Ltd., Bangalore, India

54/3020-543210 – Printed on acid-free paper

Preface

The quantum theory of magnetism is a well-developed part of contemporary solid-state physics. The basic concepts of this theory can be used to describe such important effects as ferromagnetic ordering of localized magnetic moments in crystals and ferromagnetism of metals produced by essentially delocalized electrons, as well as various types of mutual orientation of atomic magnetic moments in solids possessing different crystal lattices and compositions. In recent years,the spin-fluctuational approach has been developed, which can overcome some contradictions between "localized" and "itinerant" models in the quantum mechanics of magnetic crystals.

These are only some of the principal achievements of quantum magnetic theory. Almost all of the known magnetic properties of solids can be qualitatively explained on the basis of its concepts. Further developments should open up the possibility of reliable quantitative description of magnetic properties of solids. Unfortunately, such calculations based on model concepts appear to be very complicated and, quite often, not definite enough. The rather small number of parameters of qualitative models are usually not able to take into account the very different types of magnetic interactions that appear in crystals. Further development of magnetic theory requires quantitative information on electronic wave function in the crystal considered. This can be proved by electronic band-structure and cluster calculations. In many cases the latter can be a starting point for quantitative calculations of parameters used in magnetic theory. This is the main subject of the present book, where we consider magnetic characteristics of crystals based on the results of electronic structure calculations.

Such a quantitative theory could be the basis for theoretical modelling and a theoretical search for new magnetic alloys and compounds with interesting magnetic properties.

For more than half a century, a fundamental problem of magnetic theory has been the development of an exchange interaction model which allows the study of the nature of magnetic ordering in crystals with different types of chemical bonding: insulators, semiconductors and metals. Any reliable solution of this problem is not possible without rather accurate methods for electronic structure analysis of different crystals and their magnetic characteristics. An essential progress has been achieved recently in nonempirical calculations of the electronic energy spectrum of solids, but studies of magnetic interactions are often mainly of a model, semiphenomenological character. For magnetic insulators and semiconductors with localized electronic states, Anderson's superexchange model is often used. This model takes into account some averaged characteristics of real chemical bonding in a crystal, for example, covalency parameters. When approaching quantitative calculations of exchange interactions and the temperatures of magnetic phase transitions, detailed consideration of electronic structure and chemical bonding peculiarities becomes necessary and should be introduced into the magnetic theory.

As is known, the one-electron Stoner approach cannot describe thermodynamic properties of ferromagnetic metals. Contemporary spin-fluctuation theories of itinerant magnets are much more successful, but quite often they are based on the simple Hubbard model and do not consider carefully the real structure of electronic states in metals. Quantitative calculations of phenomenological theory parameters for real crystals appear to be complicated, not always unique problems, and make any theoretical predictions of magnetic properties of new compounds very difficult. Much more attractive are some new approaches developed recently, based essentially on the results of numerical calculations of the electronic structure of solids, which are not so extremely difficult now with the powerful computers available.

In this book, we try to analyse the current status of the theoretical modelling of magnetic properties of crystals and review the most attractive directions for further development of the magnetic theory of solid-state physics.

The book is organized as follows: Chapter 1 is an introduction; Chapters 2–4 review physical models that have been created for qualitative description of various kinds of magnetic systems. Chapter 2 considers physical models that have been designed to describe properly the properties of magnetic insulators. Chapter 3 provides a similar introduction to models describing magnetic impurities in metals. Chapter 4 is devoted to itinerant magnetic systems, especially to pure metals, in their ground state and in excited magnetic configurations. These introductory chapters provide a basis for the discussion of quantitative results obtained in numerical ab initio calculations for these systems.

Chapter 5 discusses the main lines of ab initio calculation schemes, emphasising what is especially important for the analysis of magnetic systems. Begining with the most general foundations of the spin-density functional formalism, the discussion then specifies practical methods of evaluating the Green's function and various related properties of magnetic crystals with the help of the LMTO method.

Chapter 6 discusses the calculation results available on electronic structure and magnetism of pure transition metals and their compounds and alloys. A special section is devoted to magnetic properties of high-temperature superconductors.

Chapter 7 reviews numerous calculation results obtained for magnetic impurities in simple and transition metals.

It was not our intention to give a complete bibliography of all non-empirical calculations of the electronic structure of crystals which deal with the problems of magnetic theory. Usually we discuss only the results which reveal in an obvious and consistent manner the main possibilities of computational methods in the analysis of magnetic properties of crystals. We focus on papers containing unique and distinguished results. The most important review papers connected with the problems under consideration are also cited.

We are very grateful to academician S.V. Vonsovsky for his lively interest in this book and many helpful and stimulating discussions.

Jekaterinburg, Russia
May 1992

V. A. Gubanov
A. I. Liechtenstein
A. V. Postnikov

Contents

1. Introduction

The magnetic properties of solids have a purely quantum nature. Any classical system in thermal equilibrium cannot possess a magnetic moment, even in an external magnetic field. The appearance of magnetic moments at the crystal atoms is related to the spin moments of the electrons, which can interact due to quantum exchange effects, and leads to the appearance of spontaneous magnetic ordering. The values of such exchange interactions between spin moments are defined by the character of the electron wavefunction, i.e. by chemical bonding effects.

According to their magnetic properties, all magnetic solids can be divided into two large groups. The first one includes diamagnets and Pauli paramagnets [1.1], which consist of atoms with completely occupied electronic shells. In the absence of an external magnetic field, they do not possess any magnetization. The second group is formed by compounds which reveal spontaneous magnetic order. Such solids usually include transition-metal atoms with partially field *d* and *f* electronic shells. Strong chemical bonding or delocalization of valence electrons in solids favours the formation of non-magnetic configurations with doubly occupied lowest electronic states. Almost all free atoms with partially filled *d* shells have non-zero spin and orbital moments (as follows from Hund's rule), but ony the transition metals at the end of the 3*d* row (Fe, Co, Ni) possess strong magnetic properties. Magnetization of *d* metals is defined primarily by spin moments, since the orbital magnetic moment is usually frozen due to crystal-field effects [1.1]. For the metals of the beginning of the 3d row (Sc, Ti, V), strong metallic bonding in crystals suppresses magnetic interactions completely and leads to a paramagnetic ground state of the crystal.

According to the type of magnetic ordering of magnetic moments, magnetic crystals are usually divided into collinear magnetic structures (ferromagnets, antiferromagnets, ferrimagnets) and noncollinear structures (helimagnets, speromagnets, asperomagnets [1.2]). The evolution of magnetic properties of many magnetic systems in external magnetic fields is described well by Curie–Weiss paramagnetic susceptibility [1.1]. At temperatures lower than the characteristic temperature for a crystal, i.e. the Curie temperature for ferromagnets and the Néel temperature for antiferro- and ferrimagnets, spontaneous ordering of magnetic moments takes place.

In ferromagnetic crystals, the ground-state configuration is that with parallel orientation of magnetic moments of constituent atoms. Magnetization of the crystal as a whole exists even in the absence of an external magnetic field. Typical representatives of collinear ferromagnets are transition 3*d* metals (Fe, Co, Ni), Gd, and some insulating or semiconducting oxides or chalcogenides.

The simplest collinear antiferromagnet consists of two or more independent magnetic sublattices with antiparallel orientation of spin moments, so that the total magnetization of the crystal equals zero. Some examples of antiferromagnetic crystals are Cr; Mn; ordered alloys Fe_3Mn, Pt_3Fe, etc; and insulating transition-metal compounds such as MnO, NiS, $KMnF_3$, etc [1.2].

The existence of several independent magnetic sublattices is a characteristic feature of ferrimagnetic crystals as well, but the values of the magnetic moments for different sublattices are different in the latter case. As a result, the crystal possesses a net magnetic moment. Representatives of this group of magnets are found among spinel-type compounds with complicated crystal lattices (Fe_3O_4. $NiFe_2O_4$), ferrites, etc. [1.3].

Among the non-collinear magnetic structures, the helimagnets are the best studied. In them, different orientations of spin magnetic moments of separate atoms are complemented by the existence of a long-range magnetic order with a pronounced axis of predominant magnetization. Such spiral magnetic ordering was detected, for example, in a $MnAu_2$ system and in some pure rare-earth metals [1.4]. A deviation from collinearity of the mutual orientation of the magnetic moments on the two sublattices in antiferromagnets appears as a result of anisotropic spin coupling and may be an origin of weak total spontaneous magnetization. This happens in some solids with strong spin-orbit interactions: Fe_2O_3, MnS, $CoCO_3$, etc. [1.5].

Recent progress in studies of magnetic properties of non-collinear spin structures in alloys and amorphous solids has extended the conventional classification of magnets. Speromagnets have been defined as magnetic solids in which the spin moments of the atoms have arbitrary orientations and no total magnetization nor any regular magnetic ordering between more distant than nearest neighbours occurs [1.2]. The essential difference of this structure from the usual paramagnetic state is that at temperatures lower than a definite ordering temperature the directions of the magnetic moments in a speromagnet are frozen and do not change with time. It is believed that speromagnetism appears in systems where exchange interactions between magnetic moments have almost random values. Such a situation is quite typical for many amorphous materials, and also for impurity systems such as solid solutions of magnetic ions in non-magnetic metals [1.2]. The well-known **CuMn** is a typical example of such a spin glass.

An asperomagnet is a type of magnetic solid with randomly directed spin moments which at low temperatures nevertheless exhibits an axis of predominant magnetization [1.2]. The asperomagnetic state is observed in many amorphous magnets, and it can be considered as a disordered analog of the non-collinear ferrimagnetic state.

The above classification of magnetic solids is based on the assumption that magnetic moments in crystals are more or less localized; i.e. one can speak of magnetic moments of separate atoms. The typical values of spin magnetic moments in usual localized-moment systems are several Bohr magnetons. As an example of a quite different magnetic system, intermetallic compounds $ZrZn_2$ or Sc_3In can be considered, which at low temperatures exhibit itinerant ferromagnetism, i.e. spontaneous magnetization arising from quite delocalized electronic states. The mean magnetic moment per atom is as small as 0.04–0.2 μ_B, and it cannot be attributed to any one constituent atom in particular. Rather it can be considered as an effect of an spin-density wave spontaneously formed in the crystal which is not necessarily commensurate with the lattice spacing.

As has been mentioned above, magnetic moments and hence magnetic ordering in crystals are esentially quantum mechanical. In his famous 1928 paper, *Heisenberg* [1.6] showed that the Pauli principle applied to describe the electrostatic interaction between electrons results in the appearance of an additional term in the Hamiltonian, which can be interpreted as exchange interaction. Many-electron wavefunctions should be antisymmetric with respect to interchange of spatial or spin coordinates of any pair of electrons. As a result, the energy of the system depends on the total value of its spin. Exchange interaction related to this effect appears to be much stronger than "classical" dipole-dipole interaction; namely it determines mutual orientation of magnetic moments in a crystal.

The magnetic ordering temperatures T_C (for ferromagnets) and T_N (for antiferromagnets) do not exceed 1000 K; that is essentially lower than the characteristic melting temperatures of solids. Thus chemical bonding energies are several orders of magnitude higher than the energies of exchange interactions of magnetic moments.

Different types of chemical bonding define the details of magnetic interactions in solids. Most transition–metal compounds possessing covalent bonding appear to be antiferromagnetic insulators. Exchange interaction usually increases as covalency grows. Oxides, as a rule, have higher Néel temperatures than fluorides. In non-conducting *d*-metal compounds, interatomic exchange is caused by spin dependence of covalent bonding between magnetic cation and non-magnetic anion atoms. Metallic compounds usually reveal ferromagnetic properties. Numerous phenomenological rules for determining the sign and value of exchange interaction, depending on the character of chemical bonding, have been considered [17]. As is obvious, each of them is valid for a relatively narrow group of the systems considered.

A general theory of magnetic interactions in solids should explain the reason for the existence or absence of local magnetic moments in various atoms, elucidate the nature of exchange bonding which leads to the observed type of magnetic ordering, and provide correct values of magnetic moments and magnetic ordering temperatures for any magnetic crystal. From the beginning of quantum magnetic theory, exchange interactions in solids have been considered in the framework of one of two approaches. They are the *localized electrons* approach and the *itinerant electrons* approach. The first suggests that spin moments of purely ionic origin are situated in crystal lattice sites. In the second approach the "band" itinerant electrons are considered as "correlated" by the exchange effects. As a result, the number of electrons with "up" and "down" directions of spin become unequal, and thus a magnetization of the crystal appears.

Heisenberg [1.6] investigated exchange effects in the simplest crystal consisting of N atoms and N electrons localized in non-degenerate orthogonal orbitals (one electron per atom). *Dirac* [1.8] introduced a convenient vector notation for spin interactions, describing the exchange energy of electrons by an additional term in the Hamiltonian:

$$\sum_{ij} J_{ij}(1/2 + 2\boldsymbol{S}_i\boldsymbol{S}_j), \tag{1.1}$$

where

$$J_{ij} = \int d\boldsymbol{r}\, d\boldsymbol{r}'\, \varphi_i^*(\boldsymbol{r})\varphi_j^*(\boldsymbol{r}') \frac{e^2}{|\boldsymbol{r}-\boldsymbol{r}'|} \varphi_j(\boldsymbol{r})\varphi_i(\boldsymbol{r}'). \tag{1.2}$$

Here $\boldsymbol{S}_i$, $\boldsymbol{S}_j$ are the operators of electron spin moment and J_{ij} are exchange integrals (the sum was taken over the pairs i and j).

Van Fleck [1.9] has pointed out the importance of such an exchange interaction operator for the description of magnetic properties of crystals. In the scope of the local model, the contribution to the total energy of the system depending on mutual orientation of spins can be represented by the sum of spin-spin pair interactions:

$$\boldsymbol{H}_{ex} = 2\sum_{ij} J_{ij}\boldsymbol{S}_i\boldsymbol{S}_j. \tag{1.3}$$

This expression – known as the Heisenberg–Dirac–van Fleck Hamiltonian – is usually the starting point for theoretical investigations of magnetic systems and for the discussion of experimental data available on them. The Green function method for spin operators allows derivation of a rigorous expression for the spectrum of elementary magnetic interactions (i.e. spin waves) and analysis of temperature dependence of crystal magnetization [1.10]. "Effective-field" approaches help in finding possible types of magnetic ordering in complicated crystal structures, depending on the values and signs of J_{ij} [1.11].

It should be kept in mind that the Heisenberg model has obvious limits in the study of real crystals. The most common point of view is that the exchange Hamiltonian (1.3) describes reasonably well magnetic insulators with localized d electrons and well-defined spin moments of constituent atoms. Nevertheless analysis based on the Hamiltonian (1.3) allows reproduction of observed magnetic properties of some metal-based magnetic systems as well [1.7]. Therefore, the J_{ij} values are generally considered as effective parameters that can be estimated from experimental data, and not as exchange integrals explicitly defined by (1.2). Attempts to calculate the exchange parameter for various pairs of $3d$ atoms using (1.2) resulted in the wrong order of magnitude, and quite often in the wrong sign [1.12], as compared with experimental data. For more successful calculation of exchange parameters for metal systems, peculiarities of electronic structure and chemical bonding in the crystal in question should be properly taken into account.

The other model of magnetism, the "itinerant" one, was developed by *Frenkel* [1.13]. *Bloch* [1.14], and *Stoner* [1.15]. The simplest form of this model deals with free electrons in a crystal, which have the following wavefunction: $\psi_k = \exp\{ikr\}$. Many-electron wavefunctions can be constructed as the Slater determinant of such functions, and the resulting total energy of the system is as follows:

$$\frac{E}{N} = \frac{0.535}{r_s^2}[(1+\zeta)^{5/2} + (1-\zeta)^{5/2}] - \frac{0.229}{r_s}[(1+\zeta)^{4/3} + (1-\zeta)^{4/3}]. \tag{1.4}$$

Here r_s is the radius of a sphere which contains unitary charge and $\zeta = (N_\uparrow - N_\downarrow)/(N_\uparrow + N_\downarrow)$ is the reduced magnetization [1.16].

The first term in (1.4) gives the average kinetic energy of an electron; the second term is the average exchange energy. It is commonly believed that the Coulomb energy of electrons is compensated for by their electrostatic attraction to positively charged nuclei. As follows from (1.4), for $r_s > 5.45$ (low electron density) the ferromagnetic ground state ($\zeta = 1$) has lower energy than the non-magnetic one ($\zeta = 0$). When electron correlation effects are taken into account, the interval of existence of non-magnetic solutions increases, as the critical value of r_s grows [1.17] and can reach $r_s \simeq 75 \pm 5$ [1.18]. Realistic values of electron density in metals correspond to $r_s \simeq 2$–3. Therefore, the model of "free" itinerant electrons [1.15] cannot account for ferromagnetism in transition metals.

Correct consideration of the real electronic structure of transition-metal-based systems is crucial. Their high density of states within the d band was not described in the free-electron model. Already simple quantum-mechanical considerations are able to explain why metals at the end of the $3d$ row are ferromagnets. From calculation for small groups of atoms it follows that unpaired electrons of Fe, Co, and Ni occupy the highest antibonding orbitals, which are the most localized [1.19]. This is why these elements keep their magnetic moments when forming a crystal from separate atoms.

Unfortunately, not all magnetic properties of transition metals can be adequately described by one or other of the two above-discussed models. Non-integer values of saturation magnetic moments per atom, large bonding energies, and a considerable linear contribution to the specific heat are not consistent with the Heisenberg model. On the other hand, the Curie–Weiss susceptibility of ferromagnetic metals and the observed temperature dependence of magnetization cannot be explained in terms of the itinerant electron model [1.20]. Developments of the "localized" model combined with improved description of electron-transfer effects have led to *Anderson*'s superexchange theory [1.21], which has turned out to be quite good when applied to antiferromagnetic compounds of transition metals. Deep study of correlation effects within the itinerant electron model gave rise to the spin-fluctuational theory [1.20], which explains ferromagnetic properties of transition metals. The subjects are discussed in more detail in Chaps. 2 and 4.

In general, theoretical description of magnetic solids should answer the question of the localization degree of spin states and define the parameters of interatomic magnetic coupling. The character of *d* states localization is determined by the type of chemical bonding; it can be investigated in "first-principles" electronic structure calculations for the crystals in question. Modern calculations for many-electron systems take into account the large role of correlation effects. On this basis, the type of magnetic structure and the degree of localization for magnetic *d* states can be predicted by purely theoretical means. Our aim is to show how that is done and the results to which it leads for various types of magnetic crystals.

2. Superexchange Interaction in Magnetic Insulators

Transition-metal compounds form a wide group of solids which exhibit various magnetic and electrical properties. Among them, magnetic insulating oxides are generally considered as typical examples of systems where magnetic properties are determined by strongly localized *d* states of transition-metal atoms. In the crystal lattices of such compounds, cation atoms which carry magnetic moments have non-magnetic ligand atoms as their nearest neighbours. Chemical bonding between magnetic atoms and ligands is often very strong due to large covalent mixing of metal and ligand states. The direct overlap of magnetic *d* (or *f*) orbitals of magnetic atoms appears to be practically non-existent. This is why the interaction between spin magnetic moments in insulators is completely mediated by the ligand electronic states, and this is generally discussed in terms of various superexchange theories.

This chapter considers some approaches to the description of such superexchange interactions that take into account peculiarities of covalent bond formation in the crystal.

2.1 Anderson Model of Superexchange

In 1934, *Kramers* [2.1] proposed a mechanism for exchange interaction where magnetic moments of cations were (in the framework of non-spin-dependent perturbation theory) coupled via the wavefunctions of intermediate anions. The peculiarities of such indirect exchange interaction were considered in for a simple system of four electrons at three atomic sites [2.2]. Excitations due to transfer of an electron from the anion *p* state to the partially filled *d* state were taken into account. The shortcoming of such an approach is that the magnetic interaction appears in the third and even fourth order of the perturbation theory, thus giving rise, strictly speaking, to a variety of "mechanisms" of indirect exchange [2.3, 4]. Power expansions converge very slowly in such a theory [2.5]; therefore, one mechanism transforms into the others depending on whether an orthogonal or a non-orthogonal basis set was used. Thus the results of calculations performed with only several terms in the expansion [2.6] are very doubtful.

Anderson [2.7] proposed a quite different approach to indirect exchange interaction. That gives an obvious physical picture of the effect and simplifies theoretical calculations considerably. This approach divides the description of physical properties of insulating $3d$ metal compounds into two steps. First, localized magnetic orbitals are found for unpaired electrons of the paramagnetic ion. The effect of the cation core and of (all) neighbouring diamagnetic ligands are so far taken into account, but exchange effects due to the interaction with the magnetic electrons of other centers are neglected. When a basis set like this (consisting of non-spin-dependent Wannier functions) is being constructed, Coulomb interactions between paramagnetic ions have to be incorporated into it. As a second step, exchange interactions between magnetic electrons are considered, based on the wavefunctions determined as described above. Since strong covalent mixing of metal d states and ligand orbitals is already included in the basis set, the additonal terms corresponding to the net superexchange coupling are really small, and one can effectively restrict oneself to the lowest orders of perturbation theory.

The splitting of indirect exchange interaction into two independent parts is possible because of the well-known experimental fact that the orbitals of magnetic cations do not feel magnetic interactions between the paramagnetic centers [2.5], as is clear from comparing measured induced hyperfine fields at the ligand nuclei in dilute systems with paramagnetic impurities.

Therefore, in principle, any spin configuration of the crystal can be chosen for construction of effective magnetic orbitals. For simplicity, *Anderson* took the ferromagnetic state with all the atomic spins parallel [2.7]. Using the spin-polarized Hartree–Fock method, one can obtain the Block functions $\psi_{\boldsymbol{k}}$ of such a crystal:

$$\boldsymbol{h}\psi_{\boldsymbol{k}} = \varepsilon_{\boldsymbol{k}}\psi_{\boldsymbol{k}}, \tag{2.1}$$

where $\boldsymbol{h}$ is the one-electron Hamiltonian.

Here most correlation effects such as intra-atomic Hund's exchange interaction are already accounted for when constructing one-electron states $\psi_{\boldsymbol{k}}$ for the narrow bands of transition-metal compounds [2.8]. To define localized magnetic orbitals, one has to perform only a unitary transformation to the Wannier functions:

$$\Phi_j = N^{-1/2}\sum_{\boldsymbol{k}} \psi_{\boldsymbol{k}}\, e^{i\boldsymbol{k}\boldsymbol{R}_j}. \tag{2.2}$$

Here N is the number of atoms in a crystal, and $\boldsymbol{R}_j$ are the lattice vectors. Transformation (2.2) gives a basis set equivalent to $\psi_{\boldsymbol{k}}$ only for completely occupied states [2.9]. This situation is typical for a strong ferromagnet with only one electron per magnetic atom: spin-up band is filled, the spin-down band is empty. When constructing the Φ_j functions, only the filled d band is to be taken into account, and the vacant d states should be omitted.

As the second step in the Anderson approach, let us define exchange interactions between magnetic quasiparticles described by the localized Wannier functions Φ_j. As soon as they contain a considerable fraction of ligand orbitals, they are effectively more extended in the crystal as compared with the atomic d functions. Therefore, in the description of magnetic interaction between distant cations, intermediate anions can be neglected in the first approximation. Moreover, it makes no sense to divide the exchange interaction into direct and indirect parts.

Anderson has shown [2.7] that the kinetic exchange due to virtual transfer of electrons is the main mechanism of spin coupling in insulators.

In the localized Wannier functions representation, the Hartree–Fock "band" energy operator becomes non-diagonal:

$$\hat{h} = \sum_{i\sigma} \varepsilon_i c_{i\sigma}^+ c_{i\sigma} + \sum_{i \neq j, \sigma} t_{ij} c_{i\sigma}^+ c_{j\sigma}, \tag{2.3}$$

where

$$\varepsilon_i = N^{-1/2} \sum_k \varepsilon_k; \qquad t_{ij} = N^{-1/2} \sum_k \varepsilon_k e^{ik(R_i - R_j)}.$$

Here $c_{i\sigma}^+$ and $c_{i\sigma}$ are, respectively creation and annihilation operators for electrons in the state Φ_i with spin σ; t are the transfer integrals. The kinetic exchange is a spin-dependent correction ΔE to the ground-state energy that appears in the second order (over t_{ij}) of the perturbation theory, provided the electron number is conserved [2.5]. The correction ΔE is given by the relation

$$\Delta E = -\sum_{ij\sigma\sigma'} \frac{t_{ij}^2}{U} c_{i\sigma}^+ c_{i\sigma'} c_{j\sigma'}^+ c_{j\sigma}. \tag{2.4}$$

In (2.4), U is the energy of electron transfer from one magnetic atom to the other. In the case we consider now (i.e. only one magnetic electron per atom in the crystal), the following correspondence between second quantization operators and spin moments can be established easily [2.5]:

$$\begin{aligned} &c_{i\uparrow}^+ c_{i\uparrow} + c_{i\downarrow}^+ c_{i\downarrow} = 1, \quad c_{i\uparrow}^+ c_{i\downarrow} = S_i^+, \\ &c_{i\uparrow}^+ c_{i\uparrow} - c_{i\downarrow}^+ c_{i\downarrow} = 2S_i^z, \quad c_{i\downarrow}^+ c_{i\uparrow} = S_i^-. \end{aligned} \tag{2.5}$$

The sum of the four Fermi operators over spin indices in (2.4) equals $-(1/2 + 2\boldsymbol{S}_i\boldsymbol{S}_j)$. Therefore, we come to the effective Heisenberg Hamiltonian of the form

$$\boldsymbol{H}_{ex} = \text{const} - 2 \sum_{i<j} J_{ij}^{kin} \hat{\boldsymbol{S}}_i \hat{\boldsymbol{S}}_j \tag{2.6}$$

with the antiferromagnetic kinetic exchange

$$J_{ij}^{kin} = -2t_{ij}^2/U. \tag{2.7}$$

Typical values of transfer integrals t_{ij} for insulators are ~ 0.1–0.5 eV; U equals the Coulomb correlation energy at one site, and it is of the order of 5–10 eV. Hence, the estimate of J^{kin} lies between -10 and -500 K, consistent with experimentally measured exchange interactions in most transition-metal compounds [2.5, 9].

While kinetic exchange is the main mechanism of antiferromagnetic coupling, other effects also contribute to the indirect exchange interaction parameter in further orders of the perturbation theory. Among these contributions, the conventional electrostatic exchange interaction, or ferromagnetic potential exchange, is dominant [2.5]:

$$J_{ij}^{pot} = \langle ij|ji\rangle$$

where

$$\langle ij|ji\rangle = \int d\boldsymbol{r}_1 \int d\boldsymbol{r}_2 \Phi_i^*(\boldsymbol{r}_1)\Phi_j^*(\boldsymbol{r}_2)(1/r_{12})\Phi_j(\boldsymbol{r}_1)\Phi_i(\boldsymbol{r}_2). \quad (2.8)$$

According to Anderson's estimations, the potential exchange is an order of magnitude smaller than the kinetic one, being therefore important only in the absence of electron transfer.

Although the mechanisms of indirect exchange described above are not affected by the filled ligand states Φ_l, several contributions from anion shells appear already in second order perturbation theory [2.10]. Among them is *Nesbet*'s well-known mechanism [2.11], which corresponds to the excitation of two electrons from ligand shells to the magnetic states:

$$J_{ij}^{N} = -\sum_{ll'} \frac{\langle ij|ll'\rangle\langle ll'|ij\rangle}{\Delta E_{ll'}}. \quad (2.9)$$

Such a mechanism can result in a noticeable antiferromagnetic contribution to exchange interactions when the energy of ligand-to-metal electron transfer $\Delta E_{ll'}$ is small, i.e. when energy levels of the filled non-metal band are close to that of cation magnetic states [2.12].

Another important contribution to the exchange interaction arises in the third order of perturbation theory, due to electron transfer of an empty cation state $\Phi_{j'}$ [2.7]:

$$J'_{ij} = \sum_{j'} \frac{t_{ij'}^2 I_{j'j}}{U^2}. \quad (2.10)$$

Here $I_{j'j}$ is the intra-atomic exchange integral. The mechanism for a similar ferromagnetic coupling can be very efficient, provided the electron transfer into the empty shell is strong. The I and U ratio is usually close to 0.1; therefore, the interaction mentioned above can be about 10–20% of the conventional kinetic exchange interaction.

As follows from the brief drescription given above, the second step in the Anderson approach just exploits the perturbation theory in a traditional way, with all terms having an obvious physical meaning. The major problem arising in an implementation of such a calculation scheme as a whole is confined to the first step – namely, the calculation of the wavefunction for magnetic quasiparticles. Strictly speaking, the one-electron Hamiltonian of the unrestricted Hartree–Fock method depends on the chosen spin configuration of the crystal; therefore, wavefunctions of magnetic electrons as well as transfer integrals t_{ij} are expected to be rather sensitive to the relative orientation of spin moments at i and j atomic sites.

Anderson reviews several ways to construct an optimal basis of non-spin-dependent functions of magnetic electrons [2.5], all of them reducing to one or another of the following prescriptions:

1) All the nearest magnetic atoms are replaced by non-magnetic cations.
2) d electrons at adjacent cations are excluded from consideration.
3) The spins at adjacent cations are considered to be antiparallel.
4) A ferromagnetic crystal is considered.

The first and second methods seem to be rather artificial. Moreover, they neglect the hybridization between d states of different centers and can misrepresent the electronic structure, resulting in considerable errors in transfer integrals and covalent exchange parameters.

With the "antiferromagnetic neighbours" prescription (3), one can obtain localized Hartree–Fock solutions for populated $d\uparrow$ states of the central cation, although these solutions will not be orthogonal to $d\downarrow$ states of the neighbouring cations and lead to severe problems in subsequent application of the perturbation theory. Therefore, the "ferromagnetic basis" (4) seems to be the simplest of the reasonably accurate bases.

Spin-polarized calculation for a ferromagnetic crystal produces two sets of self-consistent solutions for (2.1), namely, $\psi_{d\uparrow}(\boldsymbol{r})\cdot\alpha(\sigma)$ for spin-up and $\psi_{d\downarrow}(\boldsymbol{r})\cdot\beta(\sigma)$ for spin-down electrons. When constructing a spin-independent basis, one has to replace the coordinate part of the spin-down wavefunction with the one for the spin-up function: $\{\psi_{d\uparrow}\alpha, \psi_{d\uparrow}\beta\}$. Hence, $\psi_{d\uparrow}$ will no longer be the self-consistent solution for spin-down electrons. Nevertheless, this solution gives the orthogonal Wannier basis that is self-consistent for most electrons in the system [2.9].

To calculate the indirect exchange interaction for a complex containing a pair of magnetic cations, *Fuchikami* used the "antiferromagnetic" definition of transfer integrals [2.13], as is justified by the definition of exchange interaction in terms of energy difference between the antiferromagnetic and ferromagnetic configurations of two spins. Energy levels of the system in the ferromagnetic state are not degenerate, and electron transfer is prohibited by the Pauli principle. Therefore, kinetic exchange is possible, in principle, only for antiferromagnetic spin orientation. Thus two sets of self-consistent functions have to be constructed for both magnetic configurations. As has been pointed out [2.14], this way excludes from the very beginning the main idea of the Anderson approach (i.e. splitting the problem into two steps).

A more convenient way to define a basis of localized magnetic orbitals has been porposed in [2.10, 12]. Instead of choosing one or another spin configuration of the crystal, the average over all of them is taken. The total Hartree–Fock energy averaged over spins has the form [2.10]

$$H_0 = \sum_i n_i \langle \Phi_i | f | \Phi_j \rangle + \tfrac{1}{2} \sum_{i \neq j} n_i n_j (\langle \Phi_i \Phi_j | \Phi_i \Phi_j \rangle - \tfrac{1}{2} \langle \Phi_i \Phi_j | \Phi_j \Phi_i \rangle), \quad (2.11)$$

where f is the total one-electron operator. Equation (2.11) refers to a configuration with half the magnetic electrons having spins up and the other half spins down. Such averaging leads to the neglect of large contributions from intra-atomic Hund exchange, and is equivalent to a spin-restricted configuration of a crystal [2.8].

As *Freeman* mentions [2.12], one-center contributions from magnetic atoms to (2.11) must not be averaged; the terms with $i = j$ for singly occupied states are excluded from the sum. This leads to a difference between the equations which define doubly populated non-magnetic orbitals c_k and singly populated magnetic states a_i [2.10]:

$$hc_k = \varepsilon_k c_k + \tfrac{1}{2} \sum_k \lambda_{ik} a_i,$$

$$(h - \tfrac{1}{2} \langle a_i \| P a_i \rangle) a_i = \varepsilon_i a_i + \sum_{j \neq i} \mu_{ji} a_j + \sum_k \lambda_{ik} c_k. \quad (2.12)$$

where

$$h = f + \sum_i n_i (\langle \Phi_i \| \Phi_i \rangle - \tfrac{1}{2} \langle \Phi_i \| P \Phi_j \rangle). \quad (2.13)$$

Here $\langle \Phi \| \Phi \rangle$ and $\langle \Phi \| P \Phi \rangle$ are Coulomb and exchange operators; λ_{ik} and μ_{ji} are non-diagonal Lagrange factors.

Solution of (2.12) gives localized magnetic orbitals $\Phi_i = a_i$ [2.9]. The localization is due to the effect of a one-center exchange potential $-1/2 \langle a_i \| P a_i \rangle$, and is completely analogous to the well-known *Slater* localization of spin moments in the paramagnetic phase of crystals [2.8].

The main advantage of such a choice of zero-order Hamiltonian is that all transfer integrals between non-magnetic, completely filled states c_k; empty states e_n; and magnetic, partially filled states a_i disappear [2.10]: $t(c_m \to e_n) = t(c_m \to a_i) = t(a_i \to e_n) = 0$.

The only non-zero one-electron transfers are those between magnetic states that give rise to the strong kinetic exchange

$$t_{ij} = t(a_i \to a_j) = \langle a_i | h | a_j \rangle. \quad (2.14)$$

Here the problem of treating the transfer in the antiferromagnetic states does not arise [2.10]; contributions from filled or empty states can arise only when two-electron excitations (2.9) are considered.

Gendaira and *Tanabe* give complete expressions for different contributions to the indirect exchange up to the second order of perturbation theory [2.10].

2.2 Many-Electron Superexchange

"Magnetic" electrons in real transition-metal compounds populate degenerate orbitals, as for example the t_{2g} and e_g orbitals in a crystal field of cubic symmetry. Cation states are then described by the many-electron configuration $t_{2g}^n e_g^m$, along with the total spin S and irreducible representation of the given point symmetry group. Many-electron wavefunctions of a magnetic atom $\Psi(t_{2g}^n e_g^m SM_S \Gamma\gamma)$ are linear combinations of Slater determinants, and they are not eigenfunctions of $\boldsymbol{H}_0$ (2.7), which depends only on occupation numbers n_i [2.15]. One needs another Hamiltonian that will be invariant with respect to all point group operations. Hence all orbitals of a degenerate basis are considered to be equally populated; this corresponds to averaging over all possible terms of a given configuration [2.16]. *Stevens* discusses the importance of choosing the $\boldsymbol{H}_0$ of correct symmetry [2.17].

Now let us consider the possibility of introducing effective exchange interaction in the many-electron case [2.15–18]. The most common form of the Hamiltonian in the second quantization representation is as follows [2.19]:

$$\boldsymbol{H} = \sum_{ij}\sum_{\alpha\beta}\sum_{\sigma} \langle i\alpha|f|i\beta\rangle c_{i\alpha\sigma}^{+} c_{j\beta\sigma} + \frac{1}{2}\sum_{\substack{ij\\kl}}\sum_{\substack{\alpha\beta\\\gamma\delta}}\sum_{\sigma\sigma'} \langle i\alpha j\beta|l\delta k\gamma\rangle c_{i\alpha\sigma}^{+} c_{j\beta\sigma'}^{+} c_{k\gamma\sigma'} c_{l\delta\sigma}, \tag{2.15}$$

Here $|i\alpha\rangle$ is the complete set of localized orthonormal orbitals; indices i, j, k, l number the lattice sites and irreducible representations (t_{2g} or e_g); and α, β, γ and δ define orbital states, i.e. rows of given irreducible representations.

The zero-order Hamiltonian can be chosen in the following form [2.15]:

$$\begin{aligned}\boldsymbol{H}_0 = &\sum_i n_i \sum_{\alpha\in i} \langle\alpha|f|\alpha\rangle/d_i \\ &+ \sum_i n_i(n_i-1) \sum_{\alpha,\alpha'\in i}' [\langle\alpha\alpha'|\alpha\alpha'\rangle - \tfrac{1}{2}\langle\alpha\alpha'|\alpha'\alpha\rangle]/d_i(2d_i-1) \\ &+ \sum_{i>j} n_i n_j \sum_{\substack{\alpha\in i\\ \beta\in j}} [\langle\alpha\beta|\alpha\beta\rangle - \tfrac{1}{2}\langle\alpha\beta|\beta\alpha\rangle]/d_i d_j,\end{aligned} \tag{2.16}$$

where $n_i = \sum_{\alpha\sigma} c_{i\alpha\sigma}^{+} c_{i\alpha\sigma}$ is the "number of electrons" operator in the ith shell and d is the degeneracy of the orbital given. It is important that $\boldsymbol{H}_0$ is invariant over all point symmetry group operations since it depends on $\{n_i\}$ only. Its eigenvalues $E_0(\{n_i\})$ determine the mean energies of multiplets.

The basis of one-electron states $|i\alpha\rangle$ can be found by the variation of $\boldsymbol{H}_0$ that corresponds to the so-called hyper-Hartree–Fock method [2.8]. To define localized orbitals it is necessary to omit one-electron terms with $\alpha = \alpha'$ in the sum over these indices in (2.16), as was done in (2.11). This procedure corresponds to averaging over all possible terms of the $t_{2g}^n e_g^m$ configuration within given multiplicity S [2.20]. Note that this prescription has not so far been practically implemented [2.16].

To obtain the effective exchange Hamiltonian, we may use the operator form of *Bogolubov*'s perturbation theory [2.19]. Considering the perturbation operator as a sum of one-electron and two-electron parts [2.18] we can obtain

$$V = V_1 + V_2,$$

$$V_1 = \sum_{\substack{i\alpha \neq j\beta \\ \sigma\sigma'}} t(i\alpha \,|\, j\beta) c^+_{i\alpha\sigma} c_{j\beta\sigma'},$$

$$t(i\alpha \,|\, j\beta) = \langle i\alpha \,|\, f \,|\, j\beta \rangle + \sum_{l\delta\sigma} \langle i\alpha l\delta \,|\, j\beta l\delta \rangle (n_{l\delta} \delta_{i\alpha, l\delta});$$

$$V_2 = \frac{1}{2} \sum_{\substack{i\alpha \neq l\delta \\ j\beta \neq k\gamma \\ \sigma, \sigma'}} \langle i\alpha j\beta \,|\, l\delta k\gamma \rangle c^+_{i\alpha\sigma} c^+_{j\beta\sigma'} c_{k\gamma\sigma'} c_{l\delta\sigma}. \tag{2.17}$$

For simplicity we omit the spin-dependent part of the multielectron transition operator [2.15]. The operator form of the perturbation theory permits us to write the perturbation Hamiltonian projected onto the space of zero-order functions.

The $E_0(\{n_i\})$ states correspond to configurations with the number of electrons n_i being conserved at each atom (that is, the homeopolarity condition [2.19]). This results in simple rules for the projection operators $\boldsymbol{P}$ [2.18]:

$$\boldsymbol{P} c^+_{i_1} \dots c^+_{i_n} c_{j_1} \dots c_{j_n} \boldsymbol{P} = \begin{cases} 0 & \text{if } \{i\} = \{j\}, \\ c^+_{i_1} \dots c^+_{i_n} c_{i_1} \dots c_{i_n} & \text{if } \{i\} \neq \{j\}. \end{cases} \tag{2.18}$$

Due to such a property, the electron-trsansfer processes do not give a contribution to the energy of the system in the first order of the perturbation theory. Only in the second order do we obtain the exchange Hamiltonian of the system [2.18, 19] of the following form:

$$\hat{H}_{ex} = \sum_{i \neq j} \sum_{\substack{\alpha\beta \\ \gamma\delta}} \sum_{\sigma\sigma'} \left[\langle i\alpha j\gamma \,|\, j\delta i\beta \rangle - \frac{2t(i\alpha \,|\, j\gamma) t(j\delta \,|\, i\beta)}{U} \right]$$

$$\times c^+_{i\alpha\sigma} c_{i\beta\sigma'} c^+_{j\gamma\sigma'} c_{j\gamma\sigma}. \tag{2.19}$$

In *Bogolubov*'s perturbation theory, potential and kinetic exchange in (2.19) corresponds formally to the same order of expansion over the small parameter that is usually related to the interaction of electrons at the different sites [2.19]. In the conventional perturbation theory, some other contributions appear in the second order [2.16].

In a series of papers [2.21–24], attempts were made to revive Cramer's approach to the problem of superexchange, where the one-electron hybridization between metal and ligand states is considered to be the major perturbation. Then, the kinetic exchange comes out in the fourth order of the perturbation theory, while the potential exchange comes out in the fifth order [2.21]. The most fruitful idea proposed here is that transfers of the "anion-to-cation" type are to be considered explicitly, along with those of the "cation-to-cation" type, which are the only ones included in the simplest version of the Anderson model. In many transition-metal compounds with oxygen, selenium, and other "heavy" anions, the energy of the ligand-to-atom electron transfer can be less than the energy of the metal-to-metal transfer, thus exciting electrons from the filled anion shells.

In the Anderson theory of superexchange, the electron transitions from the filled anion states [2.10, 16] are treated in the same way. Moreover, the perturbation theory expansions converge faster in this case, since strong effects of the metal-ligand chemical bonds (i.e. the one-electron transfer effects) are already included in zero order.

There is another shortcoming in the higher orders of the perturbation theory: many new terms appear, and there is a tendency in the literature to identify them with "new" mechanisms of superexchange [2.22]. For example, an "interference" mechanism of exchange via two intermediate anions that correctly explains the non-additivity of superexchange was proposed [2.23]. In the Anderson model, however, only the transfer integral t_{ij} is additive, while the kinetic exchange includes all cross contributions as well [2.25]. Therefore, we suppose that the Hamiltonian (2.19) accounts correctly for the exchange interactions in dielectrics, provided the basis of orbitals of magnetic electrons is reasonably chosen.

A description of the exchange interaction in terms of the spin-moment operators is possible, given that $c^+_{i\alpha\sigma}$ and $\tilde{c}_{i\alpha\sigma} = (-1)^{j(\gamma)-\alpha+1/2-\sigma} c_{i(-\alpha)(-\sigma)}$ are double irreducible tensor operations of the range Γ over quasimomentum and 1/2 over spins [2.17, 18]. It is convenient to follow *Sidorov* et al. [2.18] in introducing new on-site operators:

$$\hat{W}^{sk}_{ivq} = \sum_{\substack{\alpha\beta\\ \sigma\sigma'}} \begin{bmatrix} \Gamma_\alpha \, \Gamma_\beta \, k \\ \alpha \;\; \beta \;\; q \end{bmatrix} \begin{bmatrix} 1/2 \; 1/2 \; s \\ \sigma \;\; \sigma' \;\; \nu \end{bmatrix} c^+_{i\alpha\sigma} \tilde{c}_{i\beta\sigma'} . \tag{2.20}$$

Here, [. . .] are Clebsch–Gordan coefficients. Within the $S\,\Gamma$ multiplet, the $W^{1A_1}_{iv0}$ operator can be expressed in terms of total spin $\hat{S}_\nu$, and the $W^{0T_1}_{i0q}$ operator in terms of the quasimomentum $\hat{L}_q$ [2.19]. Exchange interaction of multielectron atoms (2.19) in the most common case can be described by the expression [2.17, 18, 26]

$$H_{ex} = \sum_{ij} \sum_{\substack{sv\\ k\sigma}} J_{ij}(k_1 q_1, k_2 q_2) \hat{W}^{*sk_1}_{ivq_1} \hat{W}^{sk_2}_{jvq_2} ,$$

$$H_{ex} = -2 \sum_{i<j} J_{ij} \mathbf{S}_i \mathbf{S}_j . \tag{2.21}$$

Taking into account that the orbital degeneracy is usually removed in the crystal field, one may express the isotropic part ($k_1 = k_2 = A_1$) of this exchange interaction [2.18] for the multielectron case in the following form:

$$J_{ij} = \frac{1}{n_i n_j} \sum_{r,r'} J_{rr'} n_r n_{r'},$$

$$J_{rr'} = \sum_{\substack{\alpha \in r \\ \beta \in r'}} \left[\langle \alpha\beta | \beta\alpha \rangle - \frac{2t(\alpha|\beta)t(\beta|\alpha)}{U} \right]. \tag{2.22}$$

Here $\boldsymbol{S}_i$ is the total spin operator, and $n_i(n_r)$ is the number of unpaired electrons at the atom i (in the shell Γ).

The exchange interaction parameter for d atoms in a crystal of cubic symmetry can be expressed in terms of the contributions from different shells: $J_{e_g-e_g}$, $J_{e_g-t_{2g}}$, and $J_{t_{2g}-t_{2g}}$ [2.18, 26].

When taking into account the spin-dependent transfer, the terms corresponding to biquadratic exchange $\sim (\boldsymbol{S}_i\boldsymbol{S}_j)^2$ and to the multispin interaction $\sim (\boldsymbol{S}_i\boldsymbol{S}_j)(\boldsymbol{S}_k\boldsymbol{S}_l)$ [2.16] appear in the indirect exchange coupling operator. These non-Heisenberg exchange interactions lead to a variety of anomalous magnetic properties [2.27]. The introduction of spin-orbit interaction in the Hamiltonian (2.9) results in the Dzialoshinsky–Moriya antisymmetric exchange interaction [2.28] expressed in terms of the vector product of spins $\boldsymbol{D}_{ij}[\boldsymbol{S}_i \times \boldsymbol{S}_j]$. This interaction is important in weak antiferromagnets with non-collinear spin structures.

The situation is more complicated when conduction electrons are present in the system. Thus, for doped semiconductors one can get spin-dependent exchange coupling as $J'_{ij}\sqrt{\boldsymbol{S}_i\boldsymbol{S}_j}$ [2.29]. For a semiconductor with a fractional number of electrons per atom, thorough analysis of the effective spin interaction [2.30] shows that Bogolubov's perturbation theory is no longer applicable, since the condition of homeopolarity is not satisfied. In this case, the exchange Hamiltonian becomes a rather complicated function of spin variables.

Therefore, in the general case, the exchange interaction cannot be expressed in terms of universal spin couplings. On the contrary, they represent a part of the description of the complex multielectron structure of crystals. The correct theoretical interpretation of magnetic properties of different systems depends to a great extent on the choice of a model for exchange interactions that properly takes into account the detailed electronic structure of crystals.

2.3 Orbital Degeneracy and Magnetism

Exchange interaction for many-electron atoms can be reduced to pure spin interaction only for ions with an orbitally non-degenerate ground state. For Jahn–Teller ions such as Co^{2+} (e_g^1 configuration), V^{3+} (t_{2g}^2), orbital degrees of freedom exist as well. Orbital magnetic states can order in crystals in the same way as spin moments do, because of the cooperative Jahn–Teller effect. In magnetic insulators, spin and orbital states can be coupled via an exchange-type interaction; therefore, magnetic structure can be strongly dependent on orbital ordering [2.31].

We consider below the simplest case of doubly degenerate orbitals of e_g^1 configuration. Two orbital states belonging to the e_g representation can be considered as those with pseudospin $\tau = 1/2$, where its z projection $\tau_z = 1/2$ corresponds to the d_{z^2} function, and $\tau_z = -1/2$, to the $d_{x^2-y^2}$ function. Different orbital and spin configurations for two atoms with degenerate electronic states are shown in Fig. 2.1, where changes of energy due to kinetic exchange are also indicated for all cases [2.31]. Here Figs. 2a, b correspond to indirect exchange in the non-degenerate case; the Fig. 2.1 d configuration shows that the transfer of an electron for the ferromagnetic degenerate case is not precluded by the Pauli principle. On the contrary, kinetic exchange is maximal for this case because the transition energy diminishes due to intra-atomic Hund exchange I.

The ground state of the system discussed corresponds to a configuration where spin moments are ordered ferromagnetically, while orbital pseudospins are ordered antiferromagnetically (i.e. d_{z^2} and $d_{x^2-y^2}$ orbitals are interchanged). The complete Hamiltonian of the exchange interaction (2.19) for doubly degenerate Jahn–Teller ions can be written as follows [2.32]:

$$H_{ex} = -2\sum_{i<j} [J_1 S_i S_j + J_2 \tau_i \tau_j + 4J_3 (S_i S_j)(\tau_i \tau_j)], \tag{2.23}$$

where $J_1 = (2t^2/U)(1 - I/U)$; $J_2 = J_3 = (2t^2/U)(1 + I/U)$.

When minimizing the total energy related to the Hamiltonian (2.23), one can describe both orbital and spin structures of the crystal electronic states at the same time. On the other hand, if the orbital ordering is already known, the substitution of $\langle \tau_i \tau_j \rangle$ into (2.23) leads to the evaluation of pure spin anisotropic interactions.

When studying real compounds, one has to take into account properly the dependence of the transfer integrals $t(i\alpha|j\beta)$ on the type of orbitals α, β and on the symmetry of the spatial position of i, j sites with respect to the crystal as a whole. *Kugel* and *Khomskii* have derived the exchange Hamiltonian for a simple cubic lattice with e_g orbitals [2.32] and for degenerate t_{2g} states [2.33].

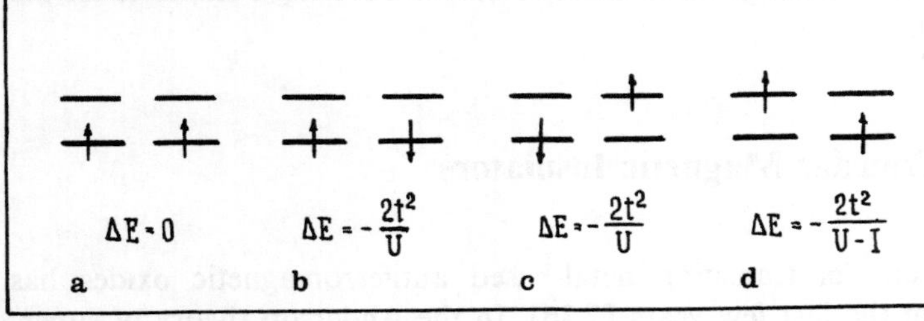

Fig. 2.1a–d. Superexchange interaction for doubly degenerate levels

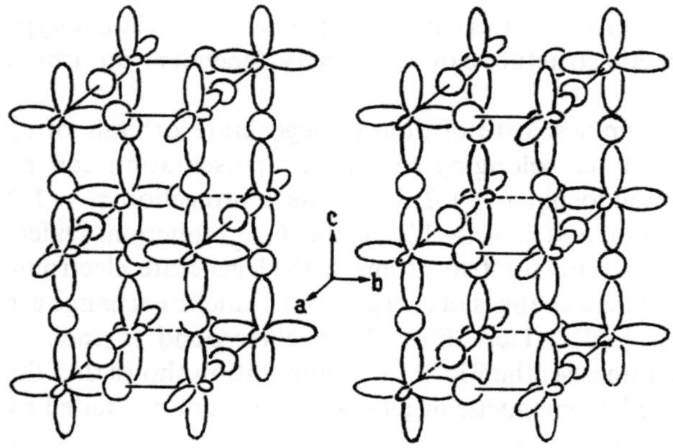

Fig. 2.2. Orbital ordering in $KCuF_3$ due to super-exchange interaction

A deeper study of the magnetic interaction (2.23) for the perovskite-type lattice [2.32] led to the conclusion that two types of orbital structure with the same type of spin ordering are possible (Fig. 2.2). Such orbital structures have been found experimentally in $KCuF_3$ [2.34]. Because of such anisotropic ordering, the exchange interactions in almost cubic $KCuF_3$ crystal are strongly anisotropic, and the crystal is actually a quasi-one-dimensional antiferromagnet. Exchange coupling parameters along the c axis are negative. They are almost an order of magnitude larger than ferromagnetic couplings in the ab plane. This result follows immediately from the theory of kinetic exchange with the exact pattern of orbitals taken into consideration: two identical half-filled orbitals interact along the c axis, giving rise to the exchange parameter $J_c = -2t^2/U$, while completely empty and completely filled orbitals overlap in the ab plane and result in $J_a = +2t^2 I/U^2$ (with $I/U \sim 0.1$) [2.31].

Analysis of the Jahn–Teller systems with t_{2g} states is more complicated because one must consider the spin-orbit interaction that can lower the orbital degeneracy [2.33]. *Eremin* and *Kalinenkov* give more rigorous and detailed descriptions of exchange interactions for many-electron Jahn–Teller ions [2.35].

To conclude this brief review of the phenomenological approach, we should mention that with the present status of such a theory it is hardly possible to predict whether the exchange mechanism or orbital ordering dominates for the compound given.

2.4 Charge-Transfer Magnetic Insulators

The understanding of transition-metal-based antiferromagnetic oxides has grown rapidly in the last few years [2.36]. In the Anderson theory of superexchange interactions, it is assumed that systems like $3d$ monoxides are

Mott–Hubbard insulators [2.37, 38]. This means that only d states are considered and that the p–d hybridization with anion states is taken into account via effective d-d hopping integrals. The charge fluctuations $d_i^n d_j^n \leftrightarrow d_i^{n+1} d_j^{n-1}$ are strongly suppressed because the involved energies are very high ($U \sim 7$–10 eV for the late transition metals). This idea provides a basis for understanding the low-lying excitations in magnetic insulators in terms of the spin-only Hamiltonian (Sect. 2.1).

There are, however, several problems with this simplest model of transition-metal compounds. Some experiments show that whereas for Ti- and V-based compounds the band gap is of the d–d type, for Co, Ni and Cu compounds the gap seems to be directly related to the electronegativity of the anion, and therefore it is probably of the charge-transfer type [2.39, 40]. It is also difficult to understand why CoS, NiS, and CuS are metals, while CoO, NiO, and CuO are insulators, since that requires the U values to be reduced to 1–2 eV in sulphide systems, compared with the 7–10 eV value typical for oxides. On the other hand, there are other many-electron excitations which do not involve the d–d correlations energy U, namely the charge-transfer fluctuations Δ: $d_i^n \leftrightarrow d_i^{n+1} L$. (Here L denotes a hole in the anion valence band.) It is also possible to consider charge fluctuations involving the $d_i^n \leftrightarrow d_i^{n-1}\, 4s$ excitations [2.41].

The model many-electron calculations of transition-metal compounds [2.36] considered anion p states as valence states, but treated the d orbitals of transition metals as impurity states and neglected the d–d bandwidth. The charge-transfer energy Δ, the effective d–d Coulomb interactions U, the width of the semielliptic anion valence band, and the hybridization interaction T between the d and p states (which are assumed to be $\boldsymbol{k}$-independent) are the parameters of this simple model. The charge-transfer parameter Δ is directly related to the electronegativity of the anion and to the Madelung potential. Both factors tend to stabilize the ionic ground state of transition-metal oxides and sulphides. The ground-state energy E_{G} of a transition-metal d^n atom in the crystals, as well as the excited-state energy of the first ionized state E_{I}^{n-1} of an $(n-1)$-electron system and electron-affinity states containing $(n+1)$ electrons with the lowest energy E_{A}^{n+1}, are calculated in the framework of the *Gunnarsson–Schoenhammer* model commonly used to describe the rare-earth materials [2.42]. The conductivity gap is given by the energy required to create an excitation in a solid in which the electron and the hole are spatially well separated and therefore uncorrelated [2.43]:

$$E_{\mathrm{gap}} = E_{\mathrm{I}}^{n-1} + E_{\mathrm{A}}^{n+1} - 2E_{\mathrm{G}}^{n}. \tag{2.24}$$

A schematic energy-level diagram for Ni^{2+} in NiO and similar compounds is shown in Fig. 2.3. The values of the energy gap obtained in this type of many-electron calculations [2.43] are shown in Fig. 2.4 as functions of Δ for various values of U. For non-zero U values, in most cases the systems are insulating ($E_{\mathrm{gap}} > 0$). For large charge-transfer energies Δ and relatively small U values, the band gap is roughly proportional to U.

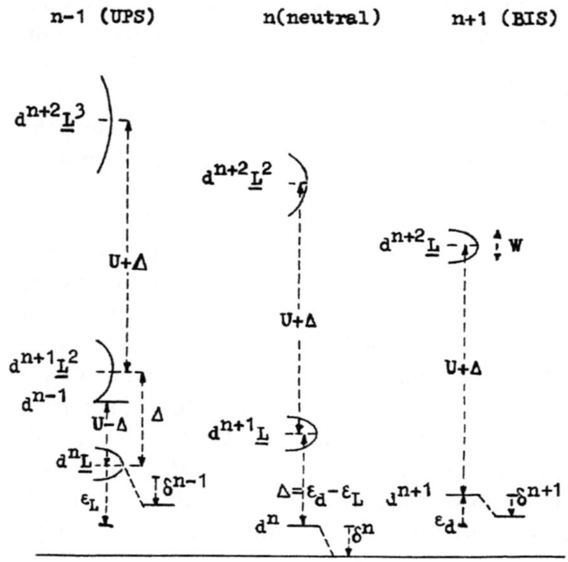

Fig. 2.3. States and parameters used in the *Zaanen–Sawatzky–Allen* theory [2.43]

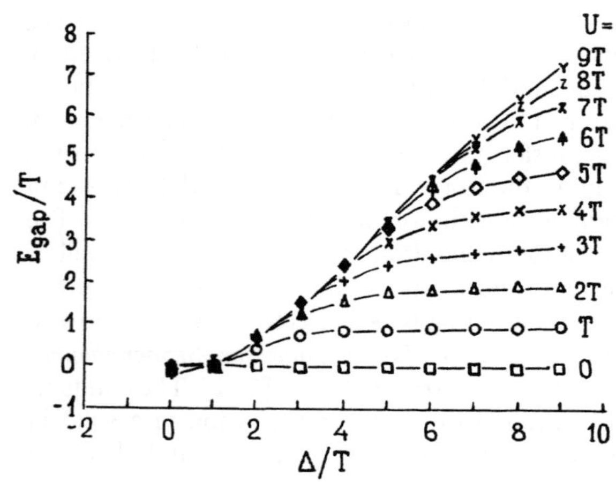

Fig. 2.4. The calculated band gap as a function of Δ for various values of U

If U exceeds Δ, the gap becomes roughly proportional to Δ (Fig. 2.4). Note that for $U = 0$ the gap is always zero, as in the simplest one-electron picture of transition-metal oxides. If the calculations take exchange interaction into account, the gap has a width proportional to the exchange splitting of the d bands.

Figure 2.5 shows the phase diagram of transition-metal compounds in the (Δ, U) plane [2.43]. Various regions have different band-gap "natures":

Region A: The case of a Mott–Hubbard insulator with $E_{gap} \sim U$; both holes

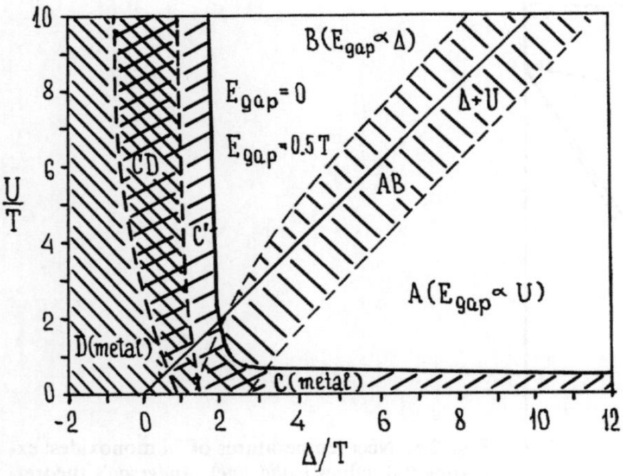

Fig. 2.5. Phase diagram of transition-metal compounds

and electrons move within the d bands. Examples are Ti_2O_3, V_2O_3 and Cr_2O_3.
Region B: Charge-transfer semiconductors with $E_{gap} \sim \Delta$ (or proportional to the electronegativity of the anion). Electrons are within the d band, and holes are within the anion valence band. Examples are CuO, $CuCl_2$ and $NiCl_2$.
Region AB (intermediate region): The hole is quite delocalized in the ligand band, but in some cases it has the symmetry properties of the d^{n-1} states of metals; electrons are in the metal d band. Examples are NiO, NiF_2 and CuF_2.

The region to the left and below the heavy solid line ($E_{gap} \sim 0.5T$) corresponds to the case when d–d hyridization and translational symmetry of the cations, which are neglected in this model, may be important, and corresponding physical systems would exhibit metallic properties.

Such a model of magnetic insulators, which includes the possibility of charge-transfer excitations, improves the simple Anderson superexchange theory; namely, it gives somewhat better estimates of the Néel temperatures for transition-metal monoxides. In Anderson's theory, where the effective hopping integrals between the states of d metals are $b = T^2/\Delta$ and $J_{ex} = -2b^2/U$, the experimentally determined values of U give wrong values of the Néel temperatures for the 3d metal oxides as compared with the experimental data (Fig. 2.6). The simplest three-center $d_i - L - d_j$ superexchange model with charge-transfer excitations was proposed by *Zaanen* and *Sawatzky* [2.36, 44], based on the configuration-interaction scheme. For the ferromagnetic spin alignment, one gets

$$|d_i d_j\rangle = |d_{i\uparrow} d_{j\uparrow}\rangle,$$

$$|dL\rangle = \frac{1}{\sqrt{2}}(|d_{i\uparrow} L_\uparrow\rangle + |L_\uparrow d_{i\uparrow}\rangle), \tag{2.25}$$

where d_i denotes the hole in the metal d states, and L refers to the p_z ligand hole. Then the energy matrix $\langle d_i|\boldsymbol{H}|L\rangle - V_\sigma$ for a ferromagnetic cluster would have

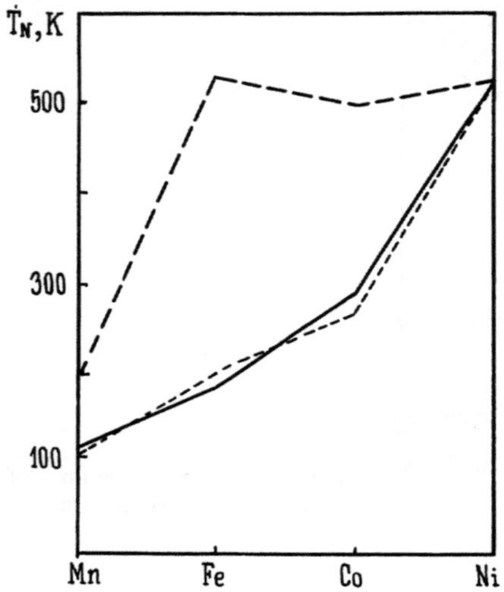

Fig. 2.6. Néel temperatures of 3*d* monoxides: experimental values (*solid line*), *Anderson*'s theoretical results (*dashed line*) [2.7], *Zaanen*'s theoretical results (*dotted line*) [2.36]

the following form:

$$\begin{matrix} |d_i d_j\rangle & |dL\rangle \end{matrix}$$

$$\begin{pmatrix} 0 & \sqrt{2}V_\sigma \\ \sqrt{2}V_\sigma & 0 \end{pmatrix}. \tag{2.26}$$

In the antiferromagnetic case, the following states can be considered:

$$\begin{aligned} |d_i\, d_j\rangle &= |d_{i\uparrow} d_{j\downarrow}\rangle, \\ |dL\rangle &= \frac{1}{\sqrt{2}}(|d_{i\uparrow} L_\downarrow\rangle + |L_\uparrow d_{j\downarrow}\rangle), \\ |L^2\rangle &= |L_\uparrow L_\downarrow\rangle, \\ |d^2\rangle &= \frac{1}{\sqrt{2}}(|d_{i\uparrow} d_{i\downarrow}\rangle + |d_{j\uparrow} d_{j\downarrow}\rangle). \end{aligned} \tag{2.27}$$

They lead to the following energy matrix:

$$\begin{matrix} |d_i d_j\rangle & |dL\rangle & |L^2\rangle & |d^2\rangle \end{matrix}$$

$$\begin{pmatrix} 0 & \sqrt{2}V_\sigma & 0 & 0 \\ \sqrt{2}V_\sigma & \Delta & \sqrt{2}V_\sigma & V_\sigma \\ 0 & \sqrt{2}V_\sigma & 2\Delta & 0 \\ 0 & V_\sigma & 0 & U \end{pmatrix}. \tag{2.28}$$

The superexchange parameter J is determined as $J = E_0(\mathrm{AF}) - E_0(\mathrm{F})$, and corresponding energies can be derived from (2.26, 28):

$$E_0(\mathrm{F}) - \frac{2V_\sigma}{E_0(\mathrm{F}) - \Delta} = 0,$$

$$E_0(\mathrm{AF}) - \left[E_0(\mathrm{AF}) - \Delta - \frac{2V_\sigma^2}{E_0(\mathrm{AF}) - 2\Delta} - \frac{V_\sigma^2}{E_0(\mathrm{AF}) - U} \right]^{-1} = 0. \qquad (2.29)$$

For Δ, $V \gg E_0(\mathrm{F, AF})$, and V_σ^2/Δ, $V_\sigma^2/U \ll 1$, these expressions result in

$$E_0(\mathrm{F}) \simeq -\frac{2V_\sigma^2}{\Delta},$$

$$E_0(\mathrm{AF}) \simeq -\frac{2V_\sigma^2}{\Delta}\left(1 + \frac{V_\sigma^2}{\Delta^2} + \frac{V_\sigma^2}{\Delta U}\right), \qquad (2.30)$$

and the effective superexchange parameter

$$J = -2b^2\left(\frac{1}{\Delta} + \frac{1}{U}\right). \qquad (2.31)$$

An estimation of the Néel temperature using this formula [2.36] results in excellent agreement with experiment (Fig. 2.6).

Although the basic physical principles behind the model seem to be reasonable, and rough estimations made according to them are promising, ab initio evaluation of various model parameters is a problem indeed. For further applications, the possibility of first-principles band-structure calculations of these parameters is especially important and is discussed in detail in Chap. 5.

3. Localized Magnetic Moments of Impurities in Metals

In this chapter, we consider magnetic properties of metallic systems. Their main characteristic is that their magnetic properties are determined by spatially extended electron states, which form conduction bands. The existence of localized magnetic moments which is quite common in $3d$ element-based magnetic insulators (Chap. 2), appears in the present case merely as one of several possible types of magnetic behaviour of the system. As a rule, the values of atomic magnetic moments in metals are smaller than those for the corresponding ions in insulators, and they do not necessarily equal an integer number of Bohr magnetons.

The simplest system, the study of which helps us to understand the peculiarities of the magnetic properties of transition metals and alloys, is the impurity (magnetic ion) embedded in a metallic host. The important interactions in this system have been studied quite comprehensively. Below we discuss on a model level basic theoretical ideas concerning the electronic spectrum peculiarities and magnetic properties of impurities. Chapter 7 gives the results of up-to-date first-principles quantitative calculations.

3.1 Virtual Bound State

When an impurity atom is embedded in a crystal, the potential felt by its electrons becomes distorted due to the effect of neighbouring atoms. The potential well usually becomes less deep, in some cases not sufficient to keep the outer electrons in truly localized states, and these electrons may enter the conduction band.

The remaining ion embedded in the sea of conduction electrons, however, affects their wavefunction and energy spectrum. This can be described in terms of scattering theory. The asymptotic wavefunction of an electron far from the scattering center, i.e. the impurity ion, can be written as

$$\chi_{\boldsymbol{k}}(\boldsymbol{r}) \simeq e^{i\boldsymbol{k}\cdot\boldsymbol{r}} + f(k, \theta)\frac{e^{ik\cdot r}}{r}, \tag{3.1}$$

where $f(k, \theta)$ is the scattering amplitude, which can be expressed in terms of phase shifts δ_l:

$$f(k,\theta) = \frac{1}{k}\sum_l [4\pi(2l+1)]^{1/2}\, e^{i\delta l}\, Y_l^0(\theta)\sin\delta_l. \tag{3.2}$$

The phase shift depends on the energy of the incoming electron; Fig. 3.1 shows the typical behaviour of $\delta_l(E)$. For δ_l close to zero or π, there is effectively no scattering. For energies close to the energy of impurity d state E_0 (which would form a bound state were it not for the effect of embedding into a crystal) $\delta_l(E)$ passes through $\pi/2$ and gives the dominating contribution to the scattering amplitude for the electrons whose energies fall within the interval w around E_0. The scattering amplitude therefore exhibits a resonance dependence on energy.

The physical meaning of the phase shift is that electrons affected by the impurity potential slow down near the impurity site and form a cloud that screens the impurity ion. Taken at infinite distance, this screening can be found through the well-known Friedel sum rule:

$$\frac{2}{\pi}\sum_l (2l+1)\delta_l(E_F) = Z. \tag{3.3}$$

Here Z is the number of electrons provided by the impurity atom to the conduction band.

This screening cloud, or the so-called virtual bound state (VBS), is a peculiar feature of a variety of metal-based impurity systems. It is characterized by the spatial distribution and by the energy spectrum of the electrons which form such a state. Although the electronic states involved in the screening process have to be expanded in a sum over l that is principle infinite, just a few terms give the main contribution to (3.2). For transition-metal impurities, which are the main subject of this chapter, the term with $l = 2$ is dominant. This can be explained as follows. Spatial charge distribution formed by the scattering of the lth partial wave is determined by the radial Schrödinger equation:

$$-\frac{1}{r^2}\frac{\partial}{\partial r}\left(r^2\frac{\partial R_l}{\partial r}\right) + \left(\frac{l(l+1)}{r^2} + V(r)\right)R_l(r,E) = ER_l(r,E). \tag{3.4}$$

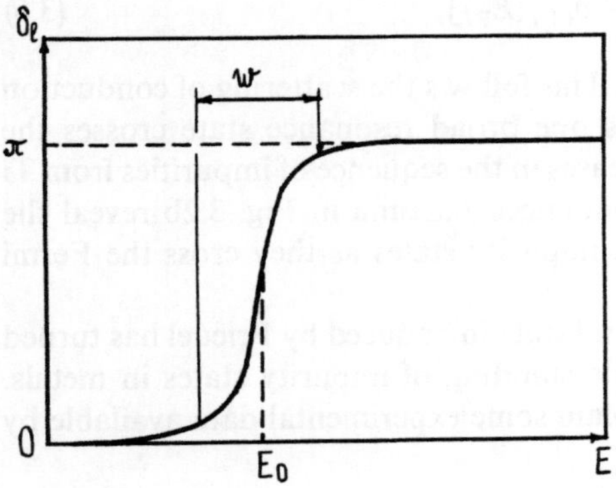

Fig. 3.1. Typical plot of the phase shift δ_l as a function of energy E

The centrifugal part of potential $l(l + 1)/r^2$ forms a potential barrier around the impurity site, which traps the electrons. For s states there is no such contribution, but it appears and increases as l grows. The average time an electron spends in the vicinity of the impurity site increases, thus, in accordance with the uncertainty principle, a narrowing of the corresponding energy level results. For electron states with $l < 2$, one can hardly speak of a pronounced resonance-like impurity state at all.

An isolated atom of a transition element with a partially filled d shell usually possesses a magnetic moment due to the intra-atomic exchange interaction. A tendency to retain a local magnetic moment remains also when the impurity is embedded in a metal; there is, however, a competing tendency to mix spin-up and spin-down states, as they are constantly exchanging their electrons via the conduction band. If the latter process involves electrons with a wide spectrum of energies within the conduction band–i.e. if the resonance state is considerably broadened–the spin-split virtual bound states may collapse into the only resonance which is degenerate over spin direction. The next section describes the conditions for such a collapse in more detail. In a simplified approach that goes back to work by *Friedel* [3.1–3], a condition for the impurity to have a local magnetic moment can be written as

$$w \leqslant p\Delta E, \tag{3.5}$$

where w is the energy width of a virtual bound state, p is the maximal possible number of unpaired d electrons in the outer shell of the impurity atom, and ΔE is the gain in energy for a pair of electrons changing their spin orientation from antiparallel to parallel. The value of $p\Delta E$ is approximately close to the magnitude of exchange splitting of the impurity d level.

Friedel's preliminary estimates of the values entering (3.5) are $\Delta E \simeq 0.6$ eV, $w \simeq 4$ eV for the impurities in Al and $w \simeq 2$ eV for the impurities in Cu [3.2]. It follows from this that no $3d$ impurities in Al possess a magnetic moment (since $p \leqslant 5$), while impurities from the middle of the $3d$ row (those withn $p \geqslant 3$) may be magnetic. This explains the change of residual resistivity in the row of $3d$ impurities in Al and Cu (Fig. 3.2). For impurities in Al, the residual resistivity is given by the formula [3.2]

$$\Delta\rho = \frac{4\pi x}{ne^2 k_F} \sum_l (l + 1) \sin^2[\delta_l(E_F) - \delta_{l+1}(E_F)], \tag{3.6}$$

where x is the impurity concentration. This follows the scattering of conduction electrons, which is mostly efficient as one broad resonance state crosses the Fermi level (for Cr impurity), but decreases in the sequence of impurities from Ti to Ni. For impurities in Cu, two pronounced maxima in Fig. 3.2b reveal the scattering at two spin-split resonance impurity states as they cross the Fermi level.

Therefore, the idea of a virtual bound state introduced by Friedel has turned out to be rather useful for better understanding of impurity states in metals. Friedel's work made it possible to explain some experimental data available by the end of the 1950s.

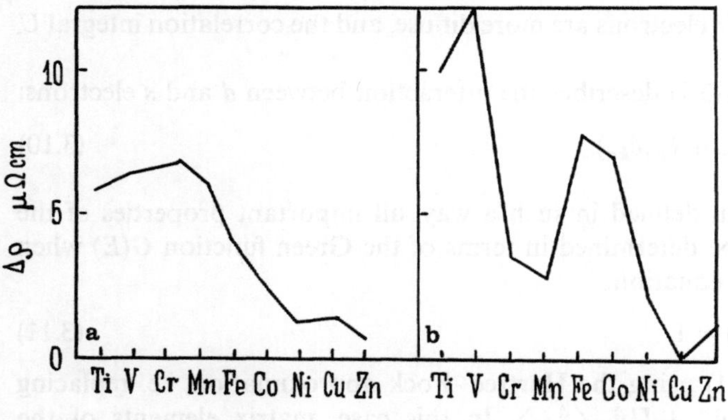

Fig. 3.2. Residual resistivity due to impurities in (a) Al and (b) Cu [3.2]

However, deeper insight into the problem was achieved only with a definite physical model which explicitly included the most important interactions. This was proposed by Anderson.

3.2 Anderson Model of Localized Magnetic Moments

Anderson's model system [3.4] included a sea of free electrons (that imitated a metal conduction band) and a finite number of localized states, or *d* states (in the original paper [3.4], only one). These states are related to the impurity atom, and the interaction between them is included, which would obviously give rise to VBS-like hybridized states.

The Hamiltonian of the Anderson model has the following form:

$$H = H_s + H_d + H_{sd}, \tag{3.7}$$

where H_s is the unperturbed energy of the system of non-interacting free or s electrons:

$$H_s = \sum_{k\sigma} E_k \hat{n}_{k\sigma}. \tag{3.8}$$

Here E_k is a one-electron energy.

In (3.7) H_d is the unperturbed energy of the impurity d state:

$$H_d = E_d(\hat{n}_{d\uparrow} + \hat{n}_{d\downarrow}) + U\hat{n}_{d\uparrow}\hat{n}_{d\downarrow},$$

$$U = \int |\varphi_d(r_1)|^2 \, |\varphi_d(r_2)|^2 \frac{e^2}{|r_1 - r_2|} dr_1 \, dr_2. \tag{3.9}$$

The second term in H_d describes the correlation between electrons and makes a difference between localized and free-electron states in this model. It is supposed that wavefunctions of s electrons are more diffuse, and the correlation integral U is negligibly small.

The third term in (3.7) describes the interaction between d and s electrons:

$$\boldsymbol{H}_{sd} = \sum_{k\sigma} V_{dk}(\hat{c}^{+}_{k\sigma}\hat{c}_{d\sigma} + \hat{c}^{+}_{d\sigma}\hat{c}_{k\sigma}). \tag{3.10}$$

With the Hamiltonian defined in such a way, all important properties of the electron system can be determined in terms of the Green function $G(E)$ when solving the following equation:

$$(E + is - \boldsymbol{H})G(E) = 1. \tag{3.11}$$

Anderson solved (3.11) using the Hartree–Fock approximation, i.e. replacing $U\hat{n}_{d\uparrow}\hat{n}_{d\downarrow}$ by $U\langle\hat{n}_{d\uparrow}\rangle\hat{n}_{d\downarrow} + U\hat{n}_{d\uparrow}\langle\hat{n}_{d\downarrow}\rangle$. In this case, matrix elements of the Hamiltonian (3.7) are easily calculated in the basis of one-electron states $\hat{c}^{+}_{d}|0\rangle$ and $\hat{c}^{+}_{k}|0\rangle$, and (3.11) reduces to the following matrix equation:

$$\begin{pmatrix} E + is - E_\sigma & -V_{dk} \\ -V_{dk'} & E + is - E_k\delta_{kk'} \end{pmatrix}\begin{pmatrix} G^{\sigma}_{dd} & G^{\sigma}_{dk} \\ G^{\sigma}_{k'd} & G^{\sigma}_{k'k} \end{pmatrix} = \begin{pmatrix} 1 & 0 \\ 0 & \delta_{kk'} \end{pmatrix}, \tag{3.12}$$

where E_σ stands for $E_d + U\langle n_{d,-\sigma}\rangle$.

All elements of the Green function can be easily derived from (3.12). First of all, of interest are the G^{σ}_{dd} elements, because they determine the density of states of d electrons interacting with the free-electron gas:

$$G^{\sigma}_{dd} = \left(E + is - E_\sigma - \sum_k \frac{|V_{dk}|^2}{E + is - E_k}\right)^{-1} \tag{3.13}$$

The sum over $\boldsymbol{k}$ in (3.13) includes the real part, which enters E_σ, and the imaginary part, which is proportional to the density of s states:

$$\lim_{s\to 0}\sum_k \frac{|V_{dk}|^2}{(E - E_k)^2 + s^2} = \langle V^2_{dk}\rangle\pi\sum_k \delta(E - E_{\mathrm{F}}) = \pi\langle V^2_{dk}\rangle n_k(E). \tag{3.14}$$

With this value labelled as Δ, we have for G^{σ}_{dd} the following relation:

$$G^{\sigma}_{dd} = \frac{1}{E - E_\sigma + i(s + \Delta)}. \tag{3.15}$$

Hence the local density of states (DOS) at the impurity site is

$$n_{d\sigma}(E) = -\frac{1}{\pi}\operatorname{Im}\{G^{\sigma}_{dd}(E)\} = \frac{1}{\pi}\frac{\Delta}{(E - E_\sigma)^2 + \Delta^2}. \tag{3.16}$$

This local DOS has the shape of a Lorentzian with half-width parameter Δ proportional to the density of electronic states in the conduction band. Integration over the local DOS (3.16) up to the Fermi energy gives the average number of electrons with given spin direction at the impurity site:

$$\langle n_{d\sigma}\rangle = \frac{1}{\pi}\operatorname{arc\,cot}\frac{E_\sigma - E_{\mathrm{F}}}{\Delta}. \tag{3.17}$$

Since $n_{d\sigma}$ depends (via E_σ) on the average number of electrons with the opposite spin direction, the actual number of electrons in both spin-split states at the impurity site can be found using the following set of self-consistent equations:

$$\langle n_{d\uparrow}\rangle = \frac{1}{\pi}\operatorname{arc\,cot}\frac{E_d - E_F + U\langle n_{d\downarrow}\rangle}{\Delta},$$

$$\langle n_{d\downarrow}\rangle = \frac{1}{\pi}\operatorname{arc\,cot}\frac{E_d - E_F + U\langle n_{d\uparrow}\rangle}{\Delta}. \tag{3.18}$$

This system can be solved numerically for any values of model parameters U, Δ, E_d, and E_F. It has either one non-magnetic or three solutions, which in the latter case include one non-magnetic and two magnetic solutions, with $n_{d\uparrow}$ and $n_{d\downarrow}$ interchanged. These two possibilities are shown in Fig. 3.3, where functions $n_{d\uparrow}(n_{d\downarrow})$ and $n_{d\downarrow}(n_{d\uparrow})$ are plotted in the same figure; crosses correspond to self-consistent solutions. If three solutions are possible (Fig. 3.3a), the non-magnetic one is usually unstable; i.e. it does not correspond to the local minimum of energy. As Fig. 3.3 shows, the condition for the curves to have three crosses – i.e. the condition for the impurity to have a local magnetic moment – is as follows:

$$\left.\frac{\partial n_\uparrow}{\partial n_\downarrow}\right|_{n_\uparrow = n_\downarrow} > -1. \tag{3.19}$$

In the field of parameters $x = (E_F - E_d)/U$. $\pi/y = \pi\Delta/U$, one can plot the critical curve that separates non-magnetic and magnetic types of solutions. The equation describing this curve follows from (3.19) when the inequality sign is replaced by an equality sign and the parameter $n = n_\uparrow = n_\downarrow$ is

$$x = n - \frac{1}{2\pi}\sin 2\pi n,$$

$$\pi/y = \sin^2 \pi n. \tag{3.20}$$

This curve, plotted in Fig. 3.4, shows that the most favourable case for the impurity to have a local magnetic moment is the case when the d shell is almost half the filled, and the width of the virtual bound state is small.

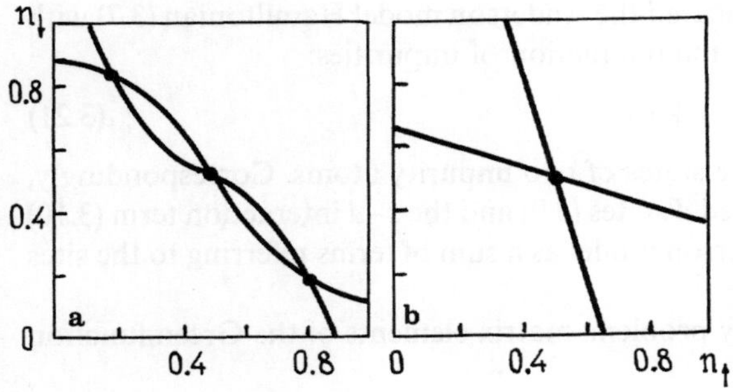

Fig. 3.3. Graphic solution of (3.18) for two sets of model parameters: **(a)** $x = 0.5$, $y = 1$; **(b)** $x = 0.5$, $y = 5$

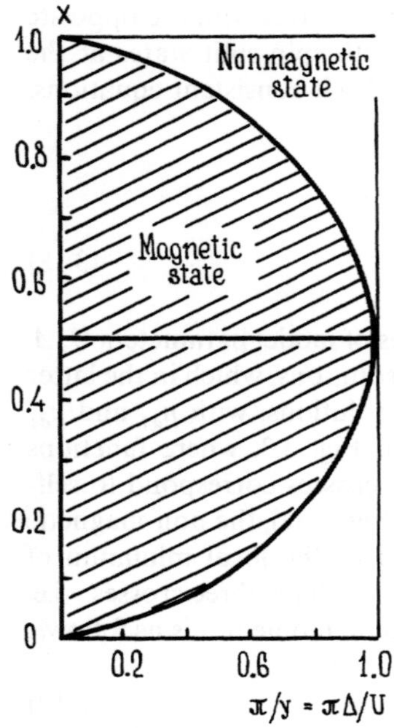

Fig. 3.4. Regions of magnetic and non-magnetic behaviour of an impurity in the field of parameters (x, π/y)

3.3 Interaction of Impurities

Alexander and *Anderson* [3.5] extended the Anderson model for the case of two closely situated impurities. *Moriya* [3.6] generalized it for orbitally degenerate states. These publications described the effects of impurity–impurity interaction on the properties of diluted alloys and allowed better understanding of important interatomic interactions in transition metals, like those that account for the type of magnetic ordering.

Alexander and *Anderson*'s model [3.5] for two equivalent impurities embedded in a free-electron gas considered the Anderson model Hamiltonian (3.7) with the following term describing the interaction of impurities:

$$\boldsymbol{H}_{12} = V_{12}(\hat{c}^{+}_{1\uparrow}\hat{c}_{2\uparrow} + \hat{c}^{+}_{1\downarrow}\hat{c}_{2\downarrow}) + \text{c.c.} \tag{3.21}$$

The indices 1 and 2 denote the states of two impurity atoms. Correspondingly, the Hamiltonian of unperturbed d states (3.9) and the s–d interaction term (3.10) appear in the Alexander-Anderson model as a sum of terms referring to the sites 1 and 2.

As with the single impurity problem, matrix elements of the Green function are determined by

$$\begin{pmatrix} E+is-E_{1\sigma} & -V_{12} & -V_{1k'} \\ -V_{21} & E+is-E_{2\sigma} & -V_{2k'} \\ -V_{k1} & -V_{k2} & (E+is-E_k)\delta_{kk'} \end{pmatrix} \begin{pmatrix} G^\sigma_{11} & G^\sigma_{12} & G^\sigma_{1k'} \\ G^\sigma_{21} & G^\sigma_{22} & G^\sigma_{2k'} \\ G^\sigma_{k1} & G^\sigma_{k2} & G^\sigma_{kk'} \end{pmatrix}$$

$$= \begin{pmatrix} 1 & 0 & 0 \\ 0 & 1 & 0 \\ 0 & 0 & \delta_{kk'} \end{pmatrix}. \tag{3.22}$$

Diagonal matrix elements of the Green function referring to one of the impurity sites are

$$G^\sigma_{11} = \left[E - E_{1\sigma} + i(s+\Delta) - \frac{V^2}{E - E_{2\sigma} + i(s+\Delta)} \right]^{-1}. \tag{3.23}$$

Here the width parameter Δ is determined as in (3.14) and is expressed as

$$V = \sum_k \frac{V_{1k}V_{k2}}{E + is - E_k} + V_{12}, \tag{3.24}$$

where V is the parameter describing the effect of direct and indirect (via conduction electrons) interaction of impurities.

After introducing new variables $n_\sigma = 1/2(n_{1\sigma} + n_{2\sigma})$, $d_\sigma = 1/2\ (n_{1\sigma} - n_{2\sigma})$ and defining the parameters $V_\sigma = \sqrt{U^2(d_\sigma)^2 + V^2}$, $E_{\sigma\pm} = E_0 + Un_{-\sigma} \pm V_{-\sigma}$. one arrives at the following formula for diagonal elements of the Green function:

$$G^\sigma_{11}(E) = \frac{1}{2}\left(\frac{1 + Ud_{-\sigma}/V_{-\sigma}}{E - E_{\sigma+} + i\Delta} + \frac{1 - Ud_{-\sigma}/V_{-\sigma}}{E - E_{\sigma-} + i\Delta} \right),$$

$$G^\sigma_{22}(E) = \frac{1}{2}\left(\frac{1 - Ud_{-\sigma}/V_{-\sigma}}{E - E_{\sigma+} + i\Delta} + \frac{1 + Ud_{-\sigma}/V_{-\sigma}}{E - E_{\sigma-} + i\Delta} \right), \tag{3.25}$$

The local DOS at any impurity site is a sum of two Lorentzians, split as a result of the interaction. Since intially the impurities were supposed to be equivalent, only the solutions corresponding to the total number of d electrons being equal at both sites are possible. There is, however, a possibility for local magnetic moments to have parallel or antiparallel mutual orientation, and these cases must be considered separately.

For parallel orientation, $n_{1\sigma} = n_{2\sigma} = n_\sigma$ and $d_\sigma = 0$, and here the local density of states following from (3.25) is

$$n_{1\sigma}(E) = n_{2\sigma}(E) = \frac{1}{2\pi}\left(\frac{\Delta}{(E - E_{\sigma+})^2 + \Delta^2} + \frac{\Delta}{(E - E_{\sigma-})^2 + \Delta^2} \right). \tag{3.26}$$

As discussed in the previous section, the condition for self-consistency can be defined as follows:

$$n_\uparrow = \frac{1}{2} + \frac{1}{2\pi}\{\arctan[y(x - n_\downarrow) - v] + \arctan[y(x - n_\downarrow) + v]\},$$

$$n_\downarrow = \frac{1}{2} + \frac{1}{2\pi}\{\arctan[y(x - n_\uparrow) - v] + \arctan[y(x - n_\uparrow) + v]\}, \qquad (3.27)$$

where $x = (E_F - E_d)/U$, $y = U/\Delta$, and $v = V/\Delta$.

The critical line in the $(x, \pi/y)$ plane separating magnetic and non-magnetic solutions is defined as follows:

$$x = n - \frac{\sin 2\pi n}{2\pi}\sqrt{v^2 \sin^2 2\pi n + 1},$$

$$\pi/y = \frac{\sin^2 2\pi n}{2\left(1 + \frac{\cos 2\pi n}{\sqrt{v^2 \sin^2 2\pi n + 1}}\right)}. \qquad (3.28)$$

Here n is a parameter.

These curves (for several values of interaction parameter v) are plotted in Fig. 3.5. For $v = 0$, (3.28) reduces to (3.20). As Fig. 3.5 shows, for $x \simeq 0.5$ (i.e. in the most favourable case for a single impurity to possess a local magnetic moment), the presence of a second impurity complicates the formation of the local moment, and higher U values are needed for the appearance of the moment. On the other hand, for x close to 0 or 1, the magnetic moment can appear at smaller U than in the case of a single impurity.

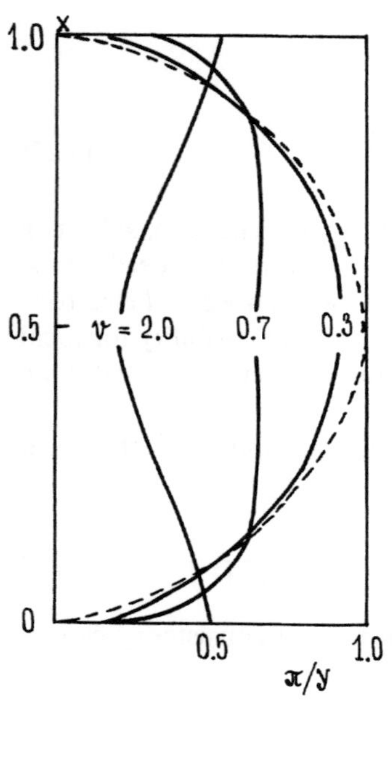

Fig. 3.5. Critical line separating regions of magnetic and non-magnetic behaviour (as in Fig. 3.4) at several values of the interaction parameter for parallel orientation of the magnetic moments of two impurities

For antiparallel orientation of local moments, $n_{1\uparrow} = n_{2\downarrow}$ and $n_{1\downarrow} = n_{2\uparrow}$, and the two independent variables are $n_\uparrow = n_\downarrow \equiv n$ and $d_\uparrow = -d_\downarrow \equiv d$. Instead of d it is convenient to use another variable, $w = \sqrt{d^2 + (V/U)^2}$, which leads to the following equations originating from (3.25):

$$n + w = \frac{1}{2} + \frac{1}{\pi} \arctan[y(x - n + w)],$$

$$n - w = \frac{1}{2} + \frac{1}{\pi} \arctan[y(x - n - w)]. \tag{3.29}$$

This set of equations seems to be similar to the condition for self-consistency (3.18) for a single impurity. But now $(n + w)$ and $(n - w)$ are the independent variables. When deriving the critical condition for antiferromagnetically ordered magnetic moments to occur, we have to keep in mind that there is a complication not found in the case of a single impurity; namely, solutions with $|w| < w_c = V/U$ are not possible. Hence the condition $w = w_c$ must be included as the additional critical condition.

By subtracting and adding the two equations (3.29), we obtain a convenient parametrization for x and y in terms of n, $w = w_c$ being an external interaction parameter:

$$x = n - \frac{w_c}{\sin 2\pi w_c} \sin 2\pi n,$$

$$\pi/y = \frac{\pi w_c}{\sin 2\pi w_c} (\cos 2\pi w_c - \cos 2\pi n). \tag{3.30}$$

For $w_c = 0$, (3.30) is reduced to (3.20). The curves determined by (3.30) are plotted in Fig. 3.6 for several values of w_c.

Note that the field of possible magnetic solutions for interaction impurities is situated inside the corresponding single impurity field for all the values of the interaction parameter. Therefore, the presence of a second impurity can only lower the tendency of magnetic moment formation.

Another difference from the case of parallel orientation is that there is a critical interaction strength at which the magnetic moments collapse completely. It happens at $V \geqslant U/2$.

When the criteria for a pair of interacting magnetic moments to exist are satisfied for both parallel and antiparallel orientations, we need to compare the energies of two magnetic configurations to find which one would really occur. It follows from the analysis [3.5] that the parallel orientation is more energetically favourable when

$$\frac{U}{2} [n_\uparrow(E_F) + n_\downarrow(E_F)] > 1, \tag{3.31}$$

i.e. when at least one of the spin-split virtual levels is situated near the Fermi level and creates a high density of states at E_F. Antiparallel orientation occurs

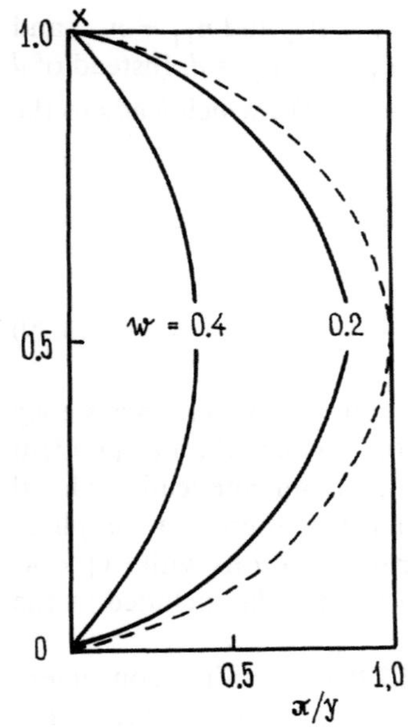

Fig. 3.6. Same as in Fig. 3.5 but for antiparallel orientation of the magnetic moments

when a virtual transition of an electron from the populated level at one site to the empty level of another site is more favourable. Because this is a virtual process, it does not need any density of states at the Fermi level.

3.4 Orbital Degeneracy and Quenching of Orbital Moment

As is shown above, the Anderson model for a single impurity with a non-degenerate d level and for a pair of interacting impurities introduces several important parameters which determine the magnetic properties of impurities. However, to describe the behaviour of real systems, one needs a more realistic model which takes into account the orbital degeneracy.

A generalization for the case of a doubly degenerate d level was considered already in *Anderson*'s initial paper [3.4]. For a more general case, the problem was solved by *Moriya* [3.6].

This generalization does not lead to particular difficulties, but the calculations involved are rather awkward. The most important changes in the initial Hamiltonian (3.7) concern the correlation term, which now includes, along with the Coulomb integrals calculated on various functions,

$$U_{\alpha\beta} = \int \varphi_\alpha^*(\boldsymbol{r}_1)\varphi_\beta^*(\boldsymbol{r}_2)\frac{e^2}{|\boldsymbol{r}_1 - \boldsymbol{r}_2|}\varphi_\alpha(\boldsymbol{r}_1)\varphi_\beta(\boldsymbol{r}_2)d\boldsymbol{r}_1\, d\boldsymbol{r}_2, \tag{3.32}$$

and also the exchange integrals

$$J_{\alpha\beta} = \int \varphi_\alpha^*(r_1)\varphi_\beta^*(r_2)\frac{e^2}{|r_1 - r_2|}\varphi_\beta(r_1)\varphi_\alpha(r_2)dr_1\,dr_2. \tag{3.33}$$

The Hamiltonian of the Anderson model then takes the form

$$H = \sum_{k,\sigma} E_k\hat{n}_{k\sigma} + \sum_{\alpha,\sigma} E_d\hat{n}_{\alpha\sigma} + \frac{1}{2}\sum_{\substack{\alpha,\beta,\sigma\\(\alpha\neq\beta)}} (U_{\alpha\beta} - J_{\alpha\beta})\hat{n}_{\alpha\sigma}\hat{n}_{\beta\sigma}$$

$$+\frac{1}{2}\sum_{\alpha,\beta,\sigma} U_{\alpha\beta}\hat{n}_{\alpha\sigma}\hat{n}_{\beta,-\sigma} + \sum_{\alpha,k,\sigma} V_{\alpha k}(\hat{c}_{k\sigma}^+\hat{c}_{\alpha\sigma} + \hat{c}_{\alpha\sigma}^+\hat{c}_{k\sigma}). \tag{3.34}$$

Here indices α and β number the orbitals and run from 1 to $(2l + 1)$. One-particle Green functions of the Hamiltonian can be found in the Hartree-Fock approximation analogously to how it was done for non-degenerate d levels (Sect. 3.2). For each orbital at the impurity site, its own density of states arises, which is a Lorentzian-shaped curve

$$n_{\alpha\sigma}(E) = \frac{1}{\pi}\frac{\Delta_\alpha}{(E - E_{\alpha\sigma})^2 + \Delta_\alpha^2} \tag{3.35}$$

with the half-width parameter $\Delta_\alpha = \pi\langle V_{\alpha k}\rangle n_\alpha(E)$. The maximum of the curve is situated at the energy

$$E_{\alpha\sigma} = E_d + U\sum_\beta \langle n_{\beta,-\sigma}\rangle + (U - J)\sum_{\beta\neq\alpha}\langle n_{\beta,\sigma}\rangle. \tag{3.36}$$

(U and J values are assumed for simplicity to be the same for all pairs of orbitals.) If one omits the α indices at Δ and assumes the peak widths to be energy independent, integration of (3.36) up to the Fermi level gives a set of $2(2l + 1)$ equations which relate the population numbers of different orbitals:

$$\Delta\cot\pi n_{\alpha,\sigma} = E_d - E_F + \sum_{\beta\neq\alpha}(U - J)n_{\beta,\sigma} + \sum_\beta U n_{\beta,-\sigma}. \tag{3.37}$$

This set of equations has a trivial solution, with all $n_{\alpha\sigma}$ equal to some value n_0. However, it is more interesting to look for the conditions when this degeneracy is lifted. This can be done by switching to new variables. $\delta n_{\alpha,\sigma} = n_{\alpha.\sigma} - n_0$, and considering resulting equation

$$-\frac{\pi\Delta\delta n_{\alpha,\sigma}}{\sin^2\pi n_0} = (U - J)\sum_{\beta\neq\alpha}\delta n_{\beta,\sigma} + U\sum_\beta \delta n_{\beta,-\sigma}, \tag{3.38}$$

which is a variation of (3.37), to have non-zero solutions. Summing up over orbitals in (3.38) leaves only two of the $2(2l + 1)$ equations, which correspond to two directions of total spin of the system:

$$-\frac{\pi\Delta}{\sin^2 \pi n_0}\delta n_\uparrow = 2l(U - J)\delta n_\uparrow + (2l + 1)U\delta n_\downarrow,$$

$$-\frac{\pi\Delta}{\sin^2 \pi n_0}\delta n_\downarrow = 2l(U - J)\delta n_\downarrow + (2l + 1)U\delta n_\uparrow, \tag{3.39}$$

where $\delta n_\sigma = \sum_\alpha \delta n_{\alpha,\sigma}$.

Subtraction of one of these equations from the other gives

$$\frac{\pi\Delta}{\sin^2 \pi n_0}(\delta n_\uparrow - \delta n_\downarrow) = (U + 2lJ)(\delta n_\uparrow - \delta n_\downarrow). \tag{3.40}$$

It can be seen that a transition from non-magnetic to magnetic behaviour of the impurity system (i.e. with $\delta n_\uparrow \neq \delta n_\downarrow$) occurs at the value of n_0 determined as

$$\frac{\pi\Delta}{\sin^2 \pi n_0} = U + 2lJ. \tag{3.41}$$

The lift of degeneracy is also possible for orbitals which belong to the same spin direction. The populations of orbitals α and β become different, and the orbital moment appears at the impurity site. One can find the corresponding critical parameters by subtracting the equations (3.38) corresponding to $\delta n_{\alpha\sigma}$ and $\delta n_{\beta\sigma}$:

$$\frac{\pi\Delta}{\sin^2 \pi n_0}(\delta n_{\alpha\sigma} - \delta n_{\beta\sigma}) = (U - J)(\delta n_{\alpha\sigma} - \delta n_{\beta\sigma}). \tag{3.42}$$

Therefore, a critical curve that separates the region of solutions with $\delta n_{\alpha\sigma} \neq \delta n_{\beta\sigma}$ is determined as follows:

$$\frac{\pi\Delta}{\sin^2 \pi n_0} = U - J. \tag{3.43}$$

To exclude n_0 from (3.41, 43), one has to solve these equations along with (3.37), setting all $n_{\alpha\sigma}$ equal to n_0. As was the case for the non-degenerate d level, the most favourable condition for the magnetic moment to occur is at $n_0 = 0.5$. In this case, the condition for the appearance of the spin moment is

$$U > \pi\Delta - 2lJ, \tag{3.44}$$

and the orbital moment appears if

$$U > \pi\Delta + J. \tag{3.45}$$

Since it is more difficult to satisfy (3.45) than (3.44), a quite common situation is for an impurity to possess the spin moment while the orbital moment is frozen. This is so in practically all cases when 3d impurities are involved.

For rare-earth impurities, the virtual bound state width Δ is smaller and the Coulomb integral U is bigger than for 3d-based systems, and the criterion (3.45) is easily satisfied. However, the ionic model [3.7] seems more appropriate than the Anderson-model-based approach in this case.

3.5 Criteria for the Existence of Magnetic Moments Based on Ab Initio Calculations

The analysis of the instability of an impurity system towards the formation of localized magnetic moments described in the previous sections is based on the Anderson model. The effects of impurity–host interaction were taken into account approximately. However, the assumption of free-electron-like behaviour for the host electronic states is by no means vital for this analysis, and quite reliable quantitative results may be obtained for actual alloy systems with the help of the Green function approach.

With this approach the criteria for single impurities, impurity pairs, or perfectly ordered magnetic crystals to carry local magnetic moments may be derived in a uniform way, based on the non-magnetic Green function $g(\boldsymbol{r}, \boldsymbol{r}', E)$ calculated self-consistently for the system given.

Spontaneous magnetic perturbation of the potential at the impurity site ΔV determines the modified Green function of the magnetic system via the Dyson equation:

$$G^\sigma = g + g\Delta V^\sigma G^\sigma . \tag{3.46}$$

The magnetic moment M resulting from this perturbation can be found as follows:

$$M = -\frac{1}{\pi}\int^{E_F} dE[n^\uparrow(E) - n^\downarrow(E)] = -\frac{1}{\pi}\operatorname{Im}\left\{\sum_\alpha \int^{E_F} dE[G^\uparrow_{\alpha\alpha}(E) - G^\downarrow_{\alpha\alpha}(E)]\right\}, \tag{3.47}$$

where $G_{\alpha\alpha}(E)$ are matrix elements of the Green function in the basis of spherical harmonics centered at the impurity site.

To find the absolute value of the self-consistent magnetic moment, (3.46, 47) should be solved self-consistently, taking into account that the magnetic moment (or rather the spatial spin distribution as a whole) determines in its turn the potential ΔV. However, if one is interested only in the critical condition for the spontaneous magnetic moment to be stable, the magnetic excitation can be treated as a first-order perturbation:

$$\Delta V^\uparrow_\downarrow = \mp\frac{IM}{2}, \tag{3.48}$$

where the value of the effective intra-atomic exchange integral I can be taken as tabulated in [3.8, 2], or calculated for any system in question along the lines described in these papers. If we want to distinguish between the "input" magnetic moment of (3.48) and the "output" magnetic moment defined by (3.47), the condition for the magnetic excitation to be self-supporting is

$$\left.\frac{\partial M_{\text{out}}}{\partial M_{\text{in}}}\right|_{M_{\text{in}}=0} > 1. \tag{3.49}$$

The condition (3.49) is practically implemented in one or another form, depending on the actual kind of magnetic excitation involved. In this simplest case of magnetic excitation at one site only, $\boldsymbol{G}^{\uparrow\downarrow} = (\boldsymbol{g}^{-1} \mp IM_{\text{in}}/2)^{-1}$, and the condition (3.49) where M_{out} is defined by (3.48) takes the form

$$I\chi_{11}(E_{\text{F}}) > 1. \tag{3.50}$$

This is the condition for an impurity (or for a given atom in a perfect crystal) to possess a local magnetic moment. The value $\chi_{11}(E_{\text{F}})$, known as the local susceptibility [3.10] is defined as follows:

$$\chi_{11}(E_{\text{F}}) = \frac{2}{\pi}\sum_{\alpha}\int^{E_{\text{F}}} \mathrm{d}E \,\mathrm{Re}\{g_{\alpha\alpha}(E)\}\,\mathrm{Im}\{g_{\alpha\alpha}(E)\}. \tag{3.51}$$

The local susceptibility can be plotted and analysed as a function of the upper integration limit in (3.51). It is essentially a positive function, which for transition metals resembles somehow the local DOS within the *d* band, with some peaks reduced or smoothed. *Terakura* et al. give examples of the $\chi_{11}(E)$ curves for some transition metals [3.10].

For the Green function of the Anderson model (3.15), the local susceptibility can be calculated analytically:

$$\chi_{11}(E) = \frac{1}{\pi}\frac{\Delta}{(E - E_{\sigma})^2 + \Delta^2}. \tag{3.52}$$

This coincides with the local DOS at the impurity site (3.16). Therefore, the criterion for the existence of a local magnetic moment in a special case of the Anderson model formally coincides with the well-known Stoner criterion for band ferromagnetism (Chap. 4). The reason for this coincidence is that local densities of states in the Anderson model are rigidly split from their initial position in the non-magnetic case when the spin splitting is introduced at the impurity site; i.e. they behave as the rigid bands of an itinerant ferromagnet. In more realistic cases when the hybridization effects of impurity *d* states with the host metal *d* band distort the impurity DOS as it acquires spin splitting, the general formula for the local susceptibility (3.51) is still valid.

The more complex situation of a pair of parallel, or antiparallel magnetic moments appearing at two impurity sites can be treated in a similar way, but with potential perturbation introduced at the two sites. The ΔV becomes a block-diagonal matrix

$$\Delta V = \left(\begin{array}{c|c} \mp\dfrac{IM}{2} & 0 \\ \hline 0 & \mp\dfrac{IM}{2} \end{array}\right), \tag{3.53}$$

each block being, in turn, a submatrix in the basis of partial states. Correspondingly, the Green function matrix consists of four blocks:

$$g = \left(\begin{array}{c|c} g_{11} & g_{12} \\ \hline g_{21} & g_{22} \end{array} \right), \tag{3.54}$$

where 1 and 2 label the sites, and the α indices are omitted for brevity. In this case, the output magnetic moment at any of the sites, say, site 1, should be expressed in terms of the input magnetic moments introduced at sites 1 and 2 simultaneously. Making use of the solution of the matrix equation (3.46),

$$G_{11} = \left\{ \left[(g_{11} - g_{12}g_{22}^{-1}g_{21})^{-1} + \frac{IM}{2} \right] - (g_{21} - g_{22}g_{12}^{-1}g_{11})^{-1} \left[(g_{22} - g_{21}g_{11}^{-1}g_{12})^{-1} + S\frac{IM}{2} \right]^{-1} (g_{12} - g_{11}g_{21}^{-1}g_{22})^{-1} \right\}^{-1}. \tag{3.55}$$

Here $S = 1$ for parallel and $S = -1$ for antiparallel orientation, and we obtain the following criterion for the existence of magnetic moments of the impurity pair:

$$I[\chi_{11}(E_F) + S\chi_{12}(E_F)] > 1, \tag{3.56}$$

where χ_{11} is the local susceptibility for site 1 defined by (3.51), and

$$\chi_{12}(E_F) = \frac{2}{\pi} \sum_{\alpha\alpha'} \int^{E_F} dE \,\mathrm{Re}\{g^{12}_{\alpha\alpha'}(E)\}\,\mathrm{Im}\{g^{12}_{\alpha\alpha'}(E)\} \tag{3.57}$$

is the non-local susceptibility. Here $\chi_{12}(E)$, taken as a function of the upper integration limit, fluctuates around zero within the d band, and it is mostly negative in the middle of the d band, if the impurities are nearest neighbours. Therefore, neighbouring magnetic impurities from the middle of the $3d$ row (Cr, Mn) exhibit antiparallel coupling of their magnetic moments (Chap. 6). In the limiting case of weak interimpurity interaction (as the distance between impurities increases), matrix elements g^{12} tend to zero, and magnetic moments become insensitive to their mutual orientation and are determined by the same criterion (3.50).

For the Green function of the Alexander–Anderson model, the equation for the critical boundary $I[\chi_{11}(E_F) + S\chi_{12}(E_F)] = 1$ is reduced to corresponding equations (3.28, 30) for the cases of parallel and antiparallel orientations of magnetic moments.

The analysis based on the concept of local and non-local susceptibilities can be useful for a much wider range of applications than problems related to impurities. First, it can be generalized to various cases of magnetic ordering in perfect crystals. For instance, the condition for a crystal to be a ferromagnet can be formulated from the "local" point of view, i.e. as a condition that equal magnetic moments appear at all sites. In analogy to the method described for two atoms, the following result can be obtained:

$$I \sum_j \chi_{1j}(E_F) > 1, \tag{3.58}$$

where the summatoion is, in principle, over all sites of the crystal around any chosen site 1. On the other hand, the criterion for ferromagnetism is known from the Stoner theory, where it follows from the assumption that spontaneous spin splitting of bands creates a self-supporting itinerant magnetic moment (Sect. 3.1):

$$In(E_F) > 1. \tag{3.59}$$

Comparison of (3.58) and (3.59) leads to the following well-known sum rule for non-local susceptibilities:

$$n(E) = \sum_j \chi_{1j}(E). \tag{3.60}$$

Note that for perfect magnetic crystals when both criteria (3.50, 59) can be applied, they are by no means equivalent. Possible situations include the following cases [3.10, 11]:

i) Local magnetic moments existing in non-ferromagnetic matter [i.e. when only (3.50) is satisfied; this corresponds, for instance, to ferromagnets above the Curie temperature].

ii) Pure itinerant ferromagnets, where magnetic moments cannot survive without ferromagnetic coupling between them [condition (3.59) but not (3.50) is satisfied; Ni is a classic example].

iii) Crystals with ferromagnetic coupling between individual spins [both (3.50) and (3.59) are satisfied], like Fe, Fe-based alloys, and magnetic insulators.

The criteria for the stability of other types of magnetic ordering in crystals can be constructed analogously to (3.56):

$$I \sum_j S_j \chi_{1j}(E_F) > 1. \tag{3.61}$$

Here the summation is again over all sites in the crystal; $S_j = 1$ if the local moment at site j is parallel to the moment at site 1 in the magnetic structure considered, and $S_j = -1$ if it is antiparallel. In such a way, several types of antiferromagnetic ordering can be considered, and those for which (3.61) is not satisfied can be excluded.

4. Exchange Interactions in Metals

Magnetic properties of ferromagnetic metals and their alloys are essentially defined by itinerant electrons described by relatively narrow energy bands. Nevertheless, thermodynamic properties of such magnetic systems can quite often be quantitatively understood within the simple Heisenberg model [4.1]. Recently substantial progress has been made in the description of the magnetic properties of transition metals. Due to correlation effects, as in the Anderson model, metal atoms can attain local magnetic moments, and the type of their ordering will be defined by exchange interactions.

This chapter discusses the early one-electron approach by Stoner and reviews current spin-fluctuation theories of itinerant magnetism.

4.1 Stoner's Model of Ferromagnetism

Most theoretical models for transition-metal magnetic properties [4.2] are based on the *Hubbard Hamiltonian*, which takes many-electron interactions at one site only [4.3]:

$$H = \sum_{ij\sigma} t_{ij} c_{i\sigma}^{+} c_{j\sigma} + U \sum_{i} n_{i\uparrow} n_{i\downarrow} \,. \tag{4.1}$$

Here t_{ij} are transfer integrals which describe the energy-band spectrum, and U is the Coulomb correlation energy at the atom given.

Equation (4.1) is actually a generalization of the Anderson model (Sect. 2.2), in which the same electron forms an energy band and reveals correlation effects. The simplest Hartree-Fock approximation for the Hamiltonian (4.1) corresponds to the one-electron Hamiltonian of the *Stoner* model [4.4, 5]:

$$H_{\mathrm{HF}} = \sum_{ij\sigma} t_{ij} c_{i\sigma}^{+} c_{j\sigma} + U \sum_{i\sigma} \langle n_{i-\sigma} \rangle n_{i\sigma} . \tag{4.2}$$

When using the set of Bloch states one obtains

$$H_{\mathrm{HF}} = \sum_{k\sigma} (E_k + I \langle N_{-\sigma} \rangle) c_{k\sigma}^{+} c_{k\sigma} . \tag{4.3}$$

Here $I = U/N_{\mathrm{at}}$ is the Stoner parameter. Thus, one-electron energies in the

Stoner model correspond to rigid shifts of energy bands for different spin directions – the shifts being in opposite directions for different spins:

$$F_{k\sigma} = F_k - IM\sigma, \tag{4.4}$$

where $M = \langle N_\uparrow \rangle - \langle N_\downarrow \rangle$ is the averaged magnetic moment.

Let us consider briefly elementary excitations and the thermodynamics of such band magnetics. In the random-phase approximation, the expression for the magnetic susceptibility of the system described by the Hamiltonian (4.2) is

$$\chi(\boldsymbol{q}, \omega) = \frac{\chi_0(\boldsymbol{q}, \omega)}{1 - I\chi_0(\boldsymbol{q}, \omega)}. \tag{4.5}$$

Here $\boldsymbol{q}$ is the wave vector, ω is the frequency, and χ_0 is one-electron susceptibility:

$$\chi_0(\boldsymbol{q}, \omega) = \sum_k \frac{n_{k\uparrow} - n_{(k+q)\downarrow}}{\hbar\omega - [E_{k\uparrow} - E_{(k+q)\downarrow}]}. \tag{4.6}$$

Elementary excitations related to characteristic features of $\chi(\boldsymbol{q}, \omega)$ in (4.5) can be of two types: first, the poles of the unenhanced susceptibility χ_0, which define the continuous band of one-electron Stoner-type excitations connected to the spin flips:

$$\hbar\omega = E_{k\uparrow} - E_{(k+q)\downarrow}. \tag{4.7}$$

Second, the condition $1 - I\chi_0 = 0$ defines the collective excitations of electron–hole pairs which are very similar to spin waves [4.6]. At small values of the wave vector, the spectrum of such excitations has the well-known spin-wave dependence [4.1]

$$\hbar\omega(\boldsymbol{q}) = D\boldsymbol{q}^2, \tag{4.8}$$

where the spin-wave stiffness constant D can be found through the metal band spectrum [4.7], using the following relation:

$$D = \frac{1}{N_\uparrow - N_\downarrow} \sum_k \left[\frac{n_{k\uparrow} + n_{k\downarrow}}{2} \nabla_k^2 E_k - \frac{N(n_{k\uparrow} - n_{k\downarrow})}{U(N_\uparrow - N_\downarrow)} |\nabla_k E_k|^2 \right]. \tag{4.9}$$

Spin waves in metals exist in a quite small long-wave region just before entering the Stoner continuum (4.7), but those many-electron excitations give the main contribution to the thermodynamic properties of itinerant magnets [4.8]. Attempts to describe the high-temperature behaviour of ferromagnetic metals in terms of Stoner's one-electron approach result in qualitatively wrong results [4.9]. For example, the temperature dependence of the magnetic susceptibility is described in the *Stoner* model by the following expression [4.5]:

$$\chi_0(T) = -\int_{-\infty}^{\infty} dE\, n(E) \frac{\partial f(E, T)}{\partial E}. \tag{4.10}$$

Here $n(E)$ is the one-electron DOS, and $f(E, T) = [\exp(E - E_F)/kT + 1]^{-1}$ is the Fermi–Dirac distribution function.

Simple estimations of the integral (4.10) lead to a quadratic dependence of the Pauli paramagnetic susceptibility on the temperature, $\chi_0 \sim T^2$, while it is known that most ferromagnetic metals exhibit the Curie-Wiess type of susceptibility $\chi \sim 1/[T - T_C]$, which is typical for systems with localized spins [4.8].

The one-electron Stoner model also fails to explain the observed values of the Curie temperatures. The temperature value at which the exchange splitting of energy bands disappears completely is proportional to the Stoner parameter I, which is of the order of 1 eV for transition metals. This value results in the Curie temperature $T_C \simeq 10\,000$ K [4.10], which exceeds experimental values by an order of magnitude. The reasons for this discrepancy are discussed in the next section. It seems that the most essential result of the one-electron Stoner model is the well-known criterion of ferromagnetism [4.5]. At $T = 0$, the condition for the metal system to be unstable with respect to spontaneous ferromagnetic band splitting immediately follows from (4.5, 10):

$$In(E_F) \geqslant 1\,, \tag{4.11}$$

where $n(E_F)$ is the total DOS at the Fermi level per atom per spin in the non-magnetic case.

When sharp singularities occur in the one-electron DOS near the Fermi level, more accurate integration in (4.10) may result in a linear dependence of χ_0 on the temperature, and thus in Curie-Weiss behaviour of the magnetic susceptibility [4.11]. Another way to improve the Hartree-Fock approximation for magnetics has been suggested [4.12, 13] for itinerant magnets with very small magnetization. In such a case, $|1 - I\chi_0| \ll 1$; therefore, additional terms connected with spin fluctuations should be taken into account in the denominator of (4.5):

$$\chi(\boldsymbol{q}, \omega) = \frac{\chi_0(\boldsymbol{q}, \omega)}{1 - I\chi_0(\boldsymbol{q}, \omega) - \lambda(T)}, \tag{4.12}$$

where the function $\lambda(T)$ is determined self-consistently [4.13]. Almost linear dependence of λ on T explains well the mechanisms of Curie–Weiss behaviour for weak ferromagnets, such as $ZrZn_2$ [4.9].

In the description of itinerant magnets with considerable magnetic moments, such as most 3d metals and their alloys, the one-electron Stoner approximation does not explain the thermodynamic properties correctly. More adequate investigation of collective spin excitations is necessary.

4.2 Spin-Fluctuation Theories of Itinerant Magnetism

Although, in principle, the Stoner model correctly describes the ground state of transition metals (at $T = 0$), the thermodynamic properties derived from it are essentially wrong. This is because the Stoner theory can treat only the evolution of magnetic moments which are translationally invariant, with their direction kept fixed. Actually, the effect of temperature is included in this case only via the Fermi step smoothing, and the estimations of the Curie temperature based on an assumption that magnetic moments drop down to zero at T_C give unreasonably high values (Sect. 4.1).

As follows from the modern view of the magnetism of metals, the thermodynamic properties of itinerant magnetis are determined primarily by many-electron excitations, e.g. spin waves [4.10]. The magnetization decreases with temperature, mostly due to the disordering of local magnetic moments. Figure 4.1 shows schematically magnetic moments varying with temperature according to two approaches.

For the discussions below, it is convenient to speak of local exchange fields at various sites. The exchange field at site i, V_i^{ex}, is determined primarily by the magnetic moment at the same site, which is a vector variable

$$V_i^{ex} = IM_i\,, \tag{4.13}$$

where I is the effective intra-atomic exchange parameter.

Spin-fluctuational approaches assume that exchange fields at non-zero temperatures vary in both magnitude and direction from site to site. The electrons are still considered to be itinerant and freely propagating in metals, and only exchange fields are localized due to modulation of uncompensated spin density on the atomic scale [4.10]. As soon as local exchange fields, in their turn, determine the spin density and hence the magnetic moment of site i, (4.13) can be considered as a self-consistency condition.

Direct implementation of this approach requires adequate inclusion of many-electron excitations in the Hubbard Hamiltonian (4.1) and therefore presents a serious problem. Actually, the only practical method for performing such calculations is the functional integral method [4.14, 15]. It allows us to reduce the two-particle problem, the Hamiltonian of which is quadratic in the charge- and spin-density operators, to the problem of one electron moving in arbitrarily fluctuating Coulomb and exchange fields. There are several ways to express the interaction term in (4.1) via charge-density and spin-density oper-

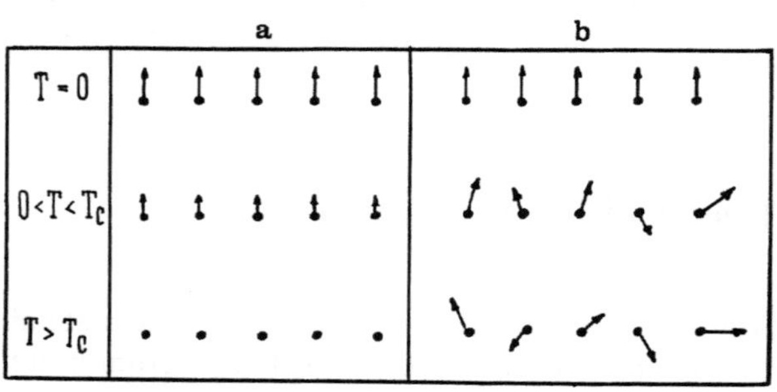

Fig. 4.1. Schematic configuration of magnetic moments in a ferromagnetic metal as assumed **(a)** the Stoner model and **(b)** Hubbard's spin-fluctuational approach [4.10]

ations. Here is the one proposed by *Wang* et al. [4.16]:

$$U\hat{n}_{i\uparrow}\hat{n}_{i\downarrow} = 1/4\,Un_i^2 - US_{iz}^2, \tag{4.14}$$

where $n_i = n_{i\uparrow} + n_{i\downarrow}$, $S_{iz} = 1/2\,(n_{i\uparrow} - n_{i\downarrow})$. In this case, only a scalar exchange fields can be introduced and that is not consistent with the rotational invariance of the spin space. As a rule, a more general representation is used, which takes into account both the ability of spin moments to deviate from the specified z direction, and the d states degeneracy in the transition-metal atoms [4.8, 10]:

$$\boldsymbol{H}_{int} = \sum_i \left[1/4\,Un_i^2 - I(\boldsymbol{S}_i\boldsymbol{e}_i)\right]. \tag{4.15}$$

Here $\boldsymbol{S}_i$ is the total spin operator at site i, and $\boldsymbol{e}_i$ is the unit vector of arbitrary direction, in respect to which an averaging has to be performed.

For a fivefold degenerate d band, $U = (9\tilde{U} - 4\tilde{I})/5$ and $I = (\tilde{U} + 4\tilde{I})/5$, where $\tilde{U}$ and $\tilde{I}$ are mean Coulomb and exchange integrals calculated within various pairs of d orbitals at the given atom [4.8].

As a result of splitting the Hubbard Hamiltonian (4.1) into a one-electron part $\boldsymbol{H}_0$ and the interaction part (4.7), one may express the statistical sum of a ferromagnet as a functional integral over exchange fields $\boldsymbol{V}$ and Coulomb fields $\boldsymbol{\Phi}$ [4.8, 10, 16, 17]:

$$Z = \int \delta\boldsymbol{V}\,\delta\boldsymbol{\Phi}\,\exp\{-\beta E[\boldsymbol{V}(t), \boldsymbol{\Phi}(t)]\}. \tag{4.16}$$

Here E is the total energy functional of a system for a given configuration of the $\boldsymbol{V}$ and $\boldsymbol{\Phi}$ fields: $\beta = (kT)^{-1}$. Strong fluctuations of charge density have low probability in the transition metals; therefore, one can neglect the Coulomb field fluctuations, as a first-order approximation [4.10]. However, the time dependence of exchange fields $\boldsymbol{V}(t)$ resulting from non-commutativity of different spin projections presents a serious problem. Up to now, no reliable methods have been developed to consider these effects properly; therefore, we confine ourselves to the static approximation, thus neglecting the time fluctuations of the exchange fields. In this case, (4.16) is transformed into a usual integral over various exchange fields, with the free energy having the form

$$E(\boldsymbol{V}) = I^{-1}\sum_i \boldsymbol{V}_i^2 + F_0(\boldsymbol{V}), \tag{4.17}$$

where $F_0(\boldsymbol{V}) = \beta^{-1}\ln\mathrm{Tr}\{\exp(-\beta\boldsymbol{H}_0 - 2\beta\sum_i \boldsymbol{V}_i\boldsymbol{S}_i)\}$ is the free energy of non-interacting electrons in exchange fields $\{\boldsymbol{V}_i\}$.

The various spin-fluctuational models differ in how they express the free energy (4.17) in arbitrary exchange fields. The simplest approximation consists of estimating the integral (4.17) by the saddle-point method [4.17]. Then the main contribution is from configurations with minimal energy $dE/d\boldsymbol{V}_i = 0$, which occurs at the following exchange field at site i [4.10, 17]:

$$\boldsymbol{V}_i^0 = -\frac{1}{2}I\left.\frac{\partial F_0}{\partial \boldsymbol{V}_i}\right|_0 = I\boldsymbol{M}_i^0. \tag{4.18}$$

Equation (4.18) is a self-consistency condition for the exchange fields (4.13) corresponding to Hartree–Fock configurations of spatially disordered magnetic moments. Therefore, in principle, the energies (4.17) can be calculated with a one-electron approach [4.18].

Hubbard [4.10] studied in detail several specific configurations of exchange fields, which elucidate the nature of magnetic excitations in ferromagnetic metals. The simplest kind of spin excitation involves change of only the z component of the exchange field at the atom in question. The change of the total energy of the whole crystal [4.10]

$$\Delta E(V_z) = I^{-1}(V_z^2 - V_{0z}^2) + \Delta F_0 . \tag{4.19}$$

As soon as the potential is changed at one site only, the free energy difference can be calculated using the scheme appropriate for solving the impurity problem (Chap. 7):

$$\Delta F_0(V_z) = -\frac{1}{\pi}\sum_{l\sigma}\int_{-\infty}^{E_F} dE\,\mathrm{Im}\{\ln[1 - \sigma V_z G_{l\sigma}(E)]\}. \tag{4.20}$$

where $G_{l\sigma}(E)$ is the Green function for the electrons in the lth subband with spin σ.

The $\Delta E(V_z)$ dependence for ferromagnetic bcc iron is shown in Fig. 4.2a [4.10]. The most interesting feature of this plot is the presence of a second

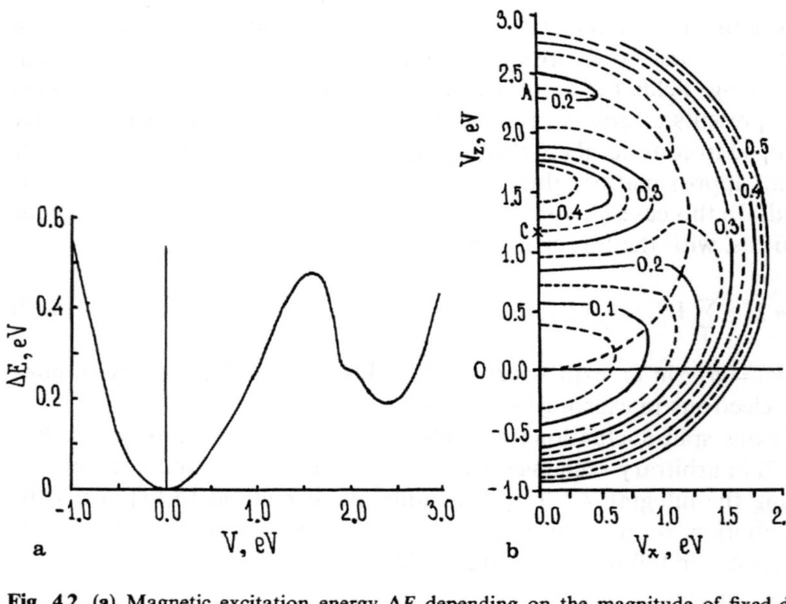

Fig. 4.2. **(a)** Magnetic excitation energy ΔE depending on the magnitude of fixed-direction exchange field V_z at one site; **(b)** Constant energy contours $\Delta E(V_z, V_x) = \text{const}$ for a deviating exchange field at one site

minimum at $V_z = 2.35$ eV. The change of the spin density at the site chosen corresponds to an almost complete inversion of the local magnetic moment. Therefore, the variation of exchange fields along the z direction only, but independently at each atom, gives quite different results from those of the Stoner model. It is more energetically favourable for the ferromagnetic system, which is excited from its $T = 0$ ground state, to invert magnetic moments at various sites independently than to lower magnetic moments gradually at all sites.

The energy of the spin flip in Fe equals 0.194 eV [4.10]. It is significantly lower than the splitting of spin-up and spin-down subbands (around 2 eV), which gives an estimate of one-electron Stoner excitations. Based on the calculated results for the energy difference, *Hubbard* [4.10] estimated the Curie temperature of Fe within the Heisenberg model, assuming $S = 1$. This estimation ($T_C = 700$ K) is in reasonable agreement with the experimental data on the magnetic transition temperature in Fe (1040 K).

More realistic excitations in ferromagnetic metals involve deviations of local exchange fields from the fixed z direction. Figure 4.2b shows the constant energy contours $E(V)$ for such spin fluctuations in iron. Low-energy excitations are almost insensitive to the specific value of the azimuthal angle; the "deviated" exchange field can be rotated around the z axis with small effect on the magnetic-moment value.

Note that these calculations describe the effect of spin fluctuations only qualitatively. If the Coulomb field fluctuations (which tend to maintain the electroneutrality of individual atoms) are taken into account, the quantitative calculations can have considerably different results [4.19].

Within the spin-fluctuational approach, the formula for effective exchange interaction in an itinerant magnet can be derived. If the free energy (4.20) is expanded over exchange fields up to second order, then, making use of (4.18), one obtains the following expression for the exchange energy:

$$E_{ex} = -\sum_{ij} J_{ij} \boldsymbol{M}_i \boldsymbol{M}_j, \tag{4.21}$$

where

$$J_{ij} = \frac{I^2 M_0^2}{2\pi} \sum_{l\sigma} \int_{-\infty}^{E_F} dE \, \mathrm{Im}\{G_{l\sigma}^{ij}(E) G_{l\sigma}^{ji}(E)\}. \tag{4.22}$$

Here $G_{l\sigma}^{ij}(E)$ is the one-electron Green function, which is calculated for equal and antiparallel magnetic moments at sites i and j: $M_i = +M_0$, $M_j = -M_0$ [4.20].

In the limit of large distance between the i and j sites, (4.21) is reduced to the well-known formula for the Ruderman-Kittel-Kasuya-Yoshida (RKKY) interaction. Note that the exchange interaction in metals can be reduced to the form typical for the Heisenberg model, with only the modification that classical mean magnetic moments, and not quantum spin operators, appear in corresponding expressions [4.21].

Possible ways to calculate thermodynamic properties of transition metals in the spin-fluctuational approach have been discussed [4.19, 22–24]. Instead of using various considerations of exchange fields, the averaging is performed over possible fluctuations of spin fields, with a weight function which depends on temperature and on the energy of these fields [4.19]. To calculate the free energy of an electron in chaotic fields, the one-site approximation within the coherent potential approximation is used, and distribution functions for exchange fields are found in a self-consistent way [4.22, 24].

The properties which specify the ferromagnetic metal in question in these calculations are one-electron density of states $n(E)$ and the Stoner exchange parameter I. In a number of actual calculations, a somewhat simplified model-type density of states is used so that the results are mostly illustrative in character. Figure 3 shows the mean magnetic moment $M_0 = 2\langle S_z \rangle$ and the average magnitude of local magnetic moment $M_l = 2(\langle S_z^2 \rangle)^{1/2}$ in a ferromagnet as functions of temperature. The $M_0(T)$ function follows a typical magnetization curve for a ferromagnet, with $M_0 = 0$ at $T = T_C$. At the same time, the local magnetic moment is only slightly reduced at higher temperatures and remains almost constant in a paramagnetic region. This result of numerical calculations [4.22–25] supports the conclusion that local magnetic moments at higher temperatures become disordered, but their magnitudes practically do not decrease.

In the above-discussed model of spin fluctuations, which is treated in the coherent-potential approximation, correlation between exchange fields at different atoms is neglected. An alternative "local band" approach [4.26–32] proceeds from the assumption that considerable short-range order is retained in a ferromagnet at temperatures above T_C. In this case, the angle between directions of exchange fields at adjacent atoms $\alpha_{ij} = (\boldsymbol{M}_i, \boldsymbol{M}_j)$ can be considered as a small parameter. The second-order perturbation theory over α_{ij} in functional integral (4.17) leads to the classical Heisenberg interaction (4.21) with the RKKY type of exchange coupling [4.30, 32]. The degree of sort-range order in ferromagnets above T_C has not been completely clarified up to now [4.33, 34]. It has been shown [4.35] that a high degree of short-range order is typical for Ni, while in Fe, spin-moment fluctuations are practically independent.

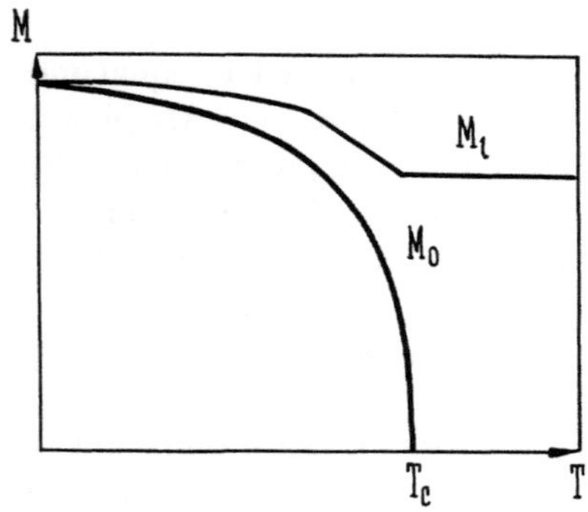

Fig. 4.3. Mean magnetic moment M_0 and local magnetic moment M_l in a ferromagnet varying with temperature [4.22]

Moriya and *Usami* [4.8, 9, 36, 37] attempted to construct a unified theory of itinerant magnetism, which is expected to be applicable to any kind of spin fluctuation. This theory is based on a phenomenological expression for F_0 (4.17) which permits description of spin excitations in both weak and strong magnetic metals and in magnetic insulators as well:

$$F_0 = -\sum_{ij} J_{ij}(M_i^2) \boldsymbol{M}_i \boldsymbol{M}_j + \Gamma(M_i^2). \tag{4.23}$$

This expression for the free energy functional takes into account both rotational degrees of freedom of exchange fields and the change of the magnetic moments themselves. Calculations of magnetic properties of itinerant systems with the phenomenological prescription for the free energy (4.23) correctly describe the whole range of thermodynamic properties for various transition-metal-based systems [4.8, 9].

The development of the spin-fluctuational models led to better understanding of transition-metal magnetism. The weak side of such theories is that in actual calculations simplified – or even model-type – density of states are usually considered, and the *d*-states degeneracy is neglected. Some examples of more realistic treatments of real magnetic systems are discussed below.

4.3 High-Temperature Magnetic Structures of Ferromagnets

An important result of spin-fluctuational theories is the finding, that low-energy excitations in ferromagnets correspond to the loss of collinearity in the local moments configuration. Rough estimation of the functional integral in saddle points leads to Hartree-Fock-type equations (4.18). Therefore, one can use one-electron approaches in studies of high-temperature magnetism. To obtain reliable quantitative results, one should take into account electronic structure and short-range order effects in a more or less straightforward way. This presents a problem within traditional band-structure approaches as soon as translational invariance of the crystal is lost in a disordered state. *You* et al. [4.18, 38] have described magnetic excitations in Fe using the tight-binding recursion method [4.39, 40].

As a starting point in the calculation performed for arbitrary magnetic configuration, local exchange fields $\boldsymbol{V}_i = V_i \boldsymbol{e}_i$ have to be introduced (unitary vector $\boldsymbol{e}$ defines the direction of the local field). The set of directions $\{\boldsymbol{e}_i\}$ can be considered as a priori given and not varying with time. Strictly speaking, arbitrary spin configuration may be involved in a precession with the frequency of the same order as the spin-wave frequency; $\hbar\omega \simeq 0.1-1$ eV. On the other hand, the frequency of electron flops from one site to another is, in the energy scale, close to the *d* bandwidth of the metal in question; i.e. $\hbar\omega \simeq 3-5$ eV. Therefore, electrons do not feel the precession, and the exchange fields can be considered as time independent in actual calculations. This is analogous to the Born–Oppenheimer approximation for atomic distortions in a crystal lattice.

In the Hartree–Fock Hamiltonian, the "band" and "exchange" parts may be specified:

$$\boldsymbol{H}_{\mathrm{HF}} = \boldsymbol{H}_{band} + \boldsymbol{H}_{ex}, \tag{4.24}$$

where the band Hamiltonian consists of the usual transfer integrals:

$$\langle im\sigma | \boldsymbol{H}_{band} | i'm'\sigma' \rangle = t_{im,i'm'} \delta_{\sigma\sigma'} . \tag{4.25}$$

The exchange part of the Hamiltonian is site-diagonal:

$$\langle im\sigma | \boldsymbol{H}_{ex} | i'm'\sigma' \rangle = -1/2\, V_i (\boldsymbol{e}_i \boldsymbol{\sigma})_{\sigma\sigma'} \delta_{ii'} \delta_{mm'} . \tag{4.26}$$

Here $\boldsymbol{\sigma}$ denotes the Pauli matrices. Therefore,

$$(\boldsymbol{e}_i \boldsymbol{\sigma})_{\sigma\sigma'} = \begin{pmatrix} \cos\theta_i & \sin\theta_i \exp(-i\varphi_i) \\ \sin\theta_i \exp(i\varphi_i) & -\cos\theta_i \end{pmatrix}, \tag{4.27}$$

where (θ, φ) are the polar and azimuthal angles of the direction $\boldsymbol{e}_i$. The exchange field V_i can be determined in a self-consistent way:

$$V_i = IM_i = I \sum_{m} \int^{E_F} [n_{im\uparrow}(E) - n_{im\downarrow}(E)]\, dE , \tag{4.28}$$

where $n_{im\sigma}(E)$ is the partial DOS for the mth orbital with spin σ at the atom i.

The local density of states in a magnetically disordered system has been calculated [4.18, 38, 41, 42] for various configurations of magnetic moments in crystal. The total energy, then, has been determined along the lines of the formula traditionally used in the Hubbard model:

$$E_{tot} = \sum_{im\sigma} \int^{E_F} E n_{im\sigma}(E)\, dE + 1/4 \sum_i IM_i^2 . \tag{4.29}$$

The second term in (4.29) introduces a correction on double-counted terms in the one-electron density of states. When considering arbitrary spin configuration, one should express the total energy in terms of exchange fields (4.28). If the self-consistency condition (4.18) is satisfied, both methods of estimating the total energy are equivalent.

The different magnetic excitations that have been considered [4.18, 38, 41, 42] are:

(a) spin spirals (SS, see Fig. 4.4): deviations of magnetic moments in the xy plane by the angle θ when going from atom to atom along the z axis;
(b) alternating tilts (AT in Fig. 4.4): magnetic moments deviate by the angles $\pm\theta$ when going along the z axis;
(c) deviation of magnetic moment at one atom;
(d) change of the angle between magnetic moments at two sites.

Figure 4.5a shows total energy and mean magnetic moment for SS and AT configurations as they were calculated by *You* and *Heine* [4.38]. The $E(\theta)$ curve for SS is everywhere lower than that for AT. For low values of θ, excitation of

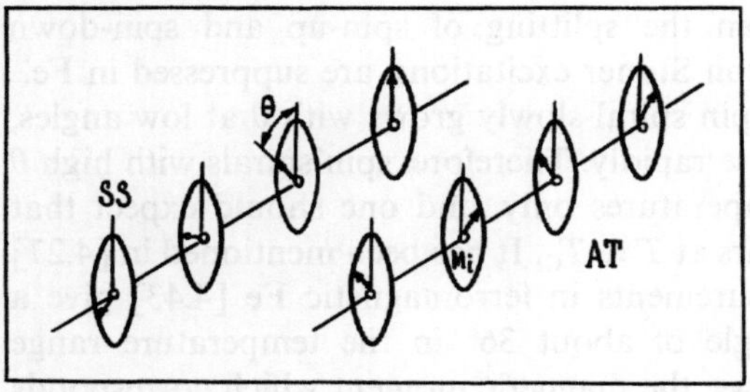

Fig. 4.4. The spatial sequence of local magnetic moments for spin spiral (SS) and alternating tilts (AT) configurations in a ferromagnet

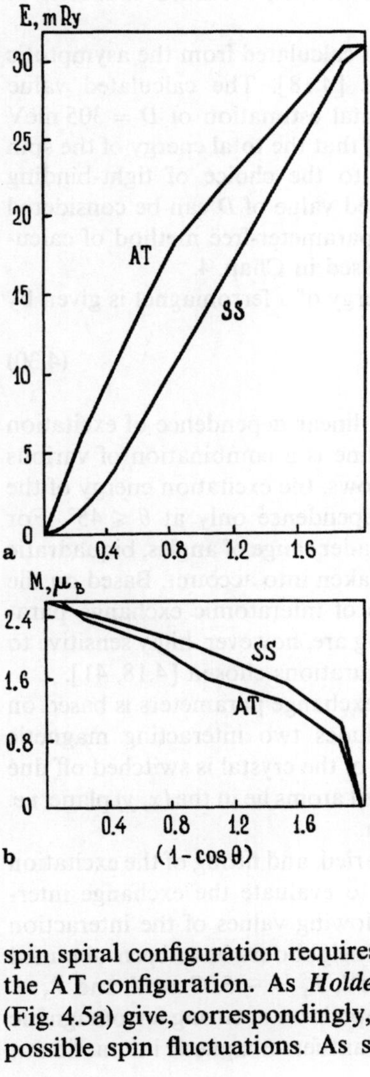

Fig. 4.5. (a) Total energy with respect to that of ferromagnetic state and **(b)** local magnetic moment varying with deviation angle θ as calculated for SS and AT configurations of bcc Fe [4.38]

spin spiral configuration requires considerably lower energy than formation of the AT configuration. As *Holden* and *You* show [4.41], E_{SS} and E_{AT} curves (Fig. 4.5a) give, correspondingly, lower and upper limits for the energies of all possible spin fluctuations. As soon as typical energies of both SS and AT configurations are far lower than the splitting of spin-up and spin-down subbands (~ 150 mRy), one-electron Stoner excitations, are suppressed in Fe.

The excitation energy of the spin spiral slowly grows with θ at low angles, and only at $\theta > 45°$ does it increase rapidly. Therefore, spin spirals with high θ angle can be excited at high temperatures only, and one should expect that strong short-range order still occurs at $T \simeq T_C$. It has been mentioned in [4.27] that the neutron scattering measurements in ferromagnetic Fe [4.43] give a mean value of the deviation angle of about 36° in the temperature range $T_C \leqslant T \leqslant 1.4T_C$. As Fig. 4.5b shows, the magnetic moment which corresponds to this angle is decreased by merely 6% as compared with the ground-state value. Therefore, magnetic moments in Fe retain their localized character at temperatures above the Curie point.

The spin-wave stiffness constant D has been calculated from the asymptotic behaviour $E_{SS} \sim D(1 - \cos\theta)$ at small angles [4.18]. The calculated value $D = 237$ meV is rather close to the experimental estimation of $D = 305$ meV [4.43]. However, it has been pointed out [4.38] that the total energy of the spin spiral at low angles becomes very sensitive to the choice of tight-binding calculation parameters. Therefore, the calculated value of D can be considered only as an order-of-magnitude estimate. The parameter-free method of calculating the spin-wave stiffness constant is discussed in Chap. 4.

In the Heisenberg model, the excitation energy of a ferromagnet is given by

$$\Delta E(\theta) = \sum_{ij} J_{ij}(1 - \cos\theta_{ij}). \tag{4.30}$$

For the AT configuration, it would result in a linear dependence of excitation energy on $(1 - \cos\theta)$, where the slope of the line is a combination of various interatomic exchange integrals. As Fig. 4.5a shows, the excitation energy of the AT configuration in Fe follows this linear dependence only at $\theta \leqslant 45°$. For correct treatment of excitation energies in a broader range of angles, biquadratic terms and even higher degrees of θ should be taken into account. Based on the $E_{SS}(\theta)$ and $E_{AT}(\theta)$ functions found, the values of interatomic exchange parameters J_{ij} can be found. The results of this fitting are, however, higly sensitive to the range of θ considered and magnetic configurations chosen [4.18, 41].

Another way of extracting the interatomic exchange parameters is based on an analysis of spin configuration which includes two interacting magnetic moments. The interaction of each with the rest of the crystal is switched off due to the fact that the magnetic moments of specified atoms lie in the (x, y) plane; i.e. they are orthogonal to the bulk magnetization.

The angle θ between two moments can be varied, and fitting of the excitation energy to the Heisenberg model enables one to evaluate the exchange interaction parameters [4.42]. For $\theta < 30°$, the following values of the interaction parameters between nearest, next-nearest, etc., neighbours have been obtained in [4.42]: $J_1 = 30.45$ meV, $J_2 = 0$, $J_3 = 5.07$ meV, $J_4 = -5.07$ meV, and $J_5 = -10.15$ meV. These values result from the competition of strong ferromagnetic coupling between nearest neighbours and weak antiferromagnetic interaction at larger distances. At $\theta \sim 60°$, the excitation energies do not obey the $(1 - \cos\theta)$ dependence, revealing that the Heisenberg model is not applicable to such kinds of magnetic interactions in iron.

A method of electronic structure calculation for the spin-spiral state which exploits the symmetry properties of this state and enables one to formulate the problem in terms of the traditional band-structure approach has been proposed by *Sandratskii* [4.44] and implemented [4.45, 46] in the study of spin-wave excitations in Fe. These calculations are not self-consistent for all the spin-spiral configurations considered, but they proceed from the self-consistent potential for the ferromagnetic ground state, and they are performed by the first-

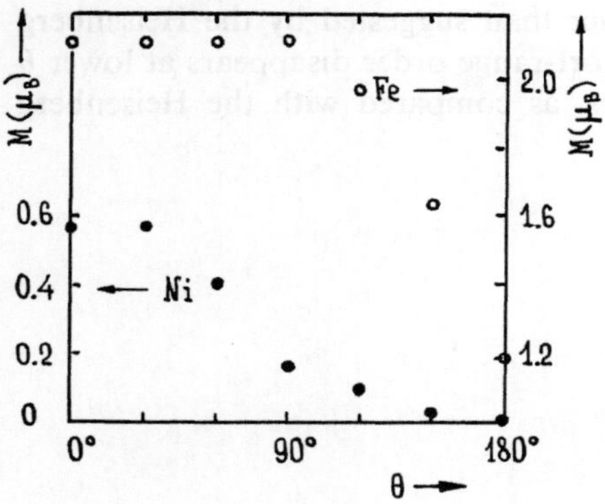

Fig. 4.6. Angular dependence of self-consistent magnetic moments in Fe and Ni. Calculation for spin spirals with wave vector $q = (0, 0, 2\pi/a)$ [4.47]

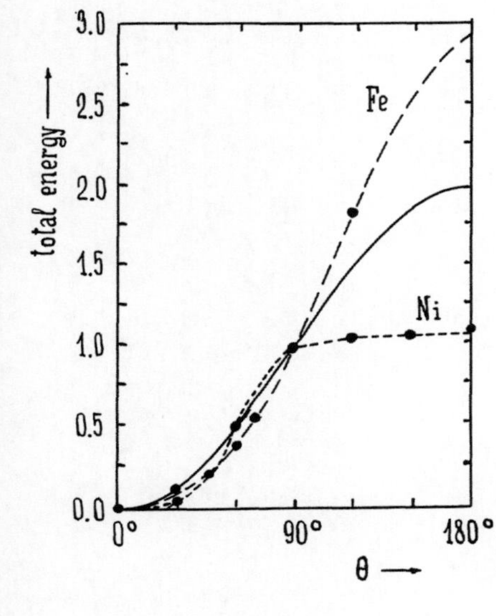

Fig. 4.7. Spin-spiral excitation energy for Fe and Ni as functions of the angle between magnetic moments at adjacent atoms. The solid curve is the $E = (1 - \cos\theta)$ line. All curves are normalized to coincide at $\theta = 90°$

principles KKR method. Therefore, their numerical results can be considered quite reliable.

Electronic structure has been calculated for spin-spiral states in Fe and Ni [4.47]. Figure 4.6 shows the evolution of the magnetic moment per atom with the θ angle. Note that the value of the local magnetic moment in iron remains practically unchanged up to $\theta = 90°$, and after a rapid decrease at $\theta > 120°$ it does not drop to zero even at $\theta = 180°$. For Ni, the magnetic moment is considerably reduced at $\theta > 30°$ and tends to zero as $\theta \to 180°$. This reveals the more itinerant nature of magnetic moments in Ni, in contrast to those of Fe.

Figure 4.7 shows the excitation energy of spin spirals as calculated with the help of (4.29) [4.47]. At $\theta < 90°$, the excitation energies of both Fe and Ni follow the Heisenberg model prescription $E \sim (1 - \cos\theta)$ with good accuracy. At larger values of θ, this agreement is violated. The fact that the excitation energy of Fe at higher θ values lies above the $(1 - \cos\theta)$ dependence reveals that the destruction of short-range order is slower than suggested by the Heisenberg model. On the other hand, for Ni the short-range order disappears at lower θ angles and hence at lower temperatures, as compared with the Heisenberg model prescription.

5. Ab Initio Approaches to the Electronic Structure of Magnetic Crystals

Up to now we have been dealing with some model approaches to the study of magnetic interactions in solids. Along with obvious advantages (clear physical picture of interactions taken into account and, in some cases, an analytical solution), such model approaches have some significant drawbacks. The drawbacks become very prominent when one tries to explain or to predict the properties of real chemical compounds. Here one encounters large uncertainties in model parameters, as well as situations where it is not possible to extract from the calculations a description of any properties which were not included in the model from the very beginning. Complete information on the electronic structure and magnetic interactions in solids can be, in principle, obtained from the solution of the Schrödinger equation for electrons which interact with other electrons and feel the external electric field created by atomic nuclei. For such a system, it is not very difficult to formulate appropriate equations and boundary conditions. What is more important, the proper choice of parameters, which presents a problem in any model approach, does not appear at all. Therefore, one may try to calculate the electronic structure of a solid starting "from first principles."

As compared with model approaches, the formulation of a quantum-mechanical problem is much simpler now, but the methods for its solution have become much more complicated. The exact solution of a many-electron Schrödinger equation for a crystal consisting of a huge number of atoms is not possible. Recent progress in the field of non-empirical "first-principles" calculations is connected with the development of computational methods which use various approximations and permit one to reduce a many-electron problem to one-particle calculations without great loss of accuracy.

5.1 Spin-Density Functional Approach

Hartree took the first step in simplifying the many-electron problem for solids and introduced the self-consistent field approximation [5.1]. With this approximation, an electron is assumed to move as an independent particle affected by the field created by the nuclei and all other electrons. The electron–electron interaction in the scope of the Hartree approximation is of purely electrostatic

(Coulomb) origin. The more sophisticated Hartree–Fock approximation [5.2] also includes the exchange interaction due to the Pauli exclusion principle. Other interactions beyond the Hartree–Fock approximation are motion correlations of electrons with the same or different spin directions. An important practical problem in the development of first-principles calculation methods is to find the model for the exchange-correlation potential which includes the effect of these correlations most accurately.

As has been proved by *Hohenberg* and *Kohn* [5.3], the ground-state total energy is a functional of the spin density:

$$E[\rho] = T_0[\rho] + \int V_{ext}(\boldsymbol{r})\rho(\boldsymbol{r})\,\mathrm{d}\boldsymbol{r} + \int \frac{\rho(\boldsymbol{r})\rho(\boldsymbol{r}')}{|\boldsymbol{r}-\boldsymbol{r}'|}\,d\boldsymbol{r}\,d\boldsymbol{r}' + E_{xc}[\rho_\uparrow, \rho_\downarrow]. \tag{5.1}$$

Here $T_0[\rho]$ is the kinetic energy functional; $V_{ext}(\boldsymbol{r})$ describes all external fields, including the Coulomb attraction to the nuclei; and E_{xc} is the exchange-correlation energy functional. It has to be specified further, and a variety of approximations have been proposed.

The important feature of the Hohenberg–Kohn theorem is that the total energy, and hence observable physical properties, do not depend on the real basis of wavefunctions used to construct the spin density. A conventional basis is presented by one-electron wavefunctions, which are determined by

$$\left(-\frac{1}{2}\Delta + V_{\mathrm{ext}}(\boldsymbol{r}) + \int \frac{\rho(\boldsymbol{r}')}{|\boldsymbol{r}-\boldsymbol{r}'|}\,d\boldsymbol{r} + V^{\sigma}_{\mathrm{xc}}(\boldsymbol{r})\right)\psi_{i\sigma}(\boldsymbol{r}) = \varepsilon_{i\sigma}\psi_{i\sigma}(\boldsymbol{r}). \tag{5.2}$$

where σ denotes "up" or "down" spin direction. The spin density then is

$$\rho^{\sigma}(\boldsymbol{r}) = \sum_i |\psi_{i\sigma}(\boldsymbol{r})|^2. \tag{5.3}$$

The exchange-correlation potential $V^{\sigma}_{\mathrm{xc}}(\boldsymbol{r})$ in (5.2) is the functional derivative of the exchange-correlation energy functional $E_{xc}[\rho_\uparrow, \rho_\downarrow]$ over the spin density $\rho_\sigma(\boldsymbol{r})$. It follows from the minimum condition for the total energy (5.1).

Since the exact expression for $E_{xc}[\rho_\uparrow, \rho_\downarrow]$ is not known, some approximations have to be used in practical calculations. An assumption that the exchange-correlation energy at any point depends on the spin density at the same point only (supposing that the density is of the same value elsewhere, i.e. it is uniform) results in

$$E_{xc}[\rho_\uparrow, \rho_\downarrow] = \int d\boldsymbol{r}\,\rho(\boldsymbol{r})\varepsilon_{xc}(\rho_\uparrow(\boldsymbol{r}), \rho_\downarrow(\boldsymbol{r})). \tag{5.4}$$

This expression is known as the local spin-density approximation (LSDA). Here ε_{xc} is the exchange-correlation energy per particle. It contains a purely exchange part ε_x due to the interaction of an electron with the "Fermi hole", i.e. the region around a given electron from which the "excess" unitary charge of the same spin direction is excluded, according to the Pauli principle.

At first sight, the assumption that ε_{xc} is a function and not a functional of the local spin density seems to be rather crude. However, only the interaction with the spherically averaged part of the Fermi hole gives a contribution to the energy, and spatial fluctuations of the exchange charge density are not important. The spherically averaged part of $\rho_x(\boldsymbol{r}, \boldsymbol{r}')$ can be described within the local spin-density approximation fairly well [5.4].

The charge density at the Fermi hole centred at $\boldsymbol{r}$ is therefore

$$\rho_x^\sigma(\boldsymbol{r}, \boldsymbol{r}') = -\rho^\sigma(\boldsymbol{r}) f(|\boldsymbol{r} - \boldsymbol{r}'|), \tag{5.5}$$

the radial distribution function being normalized as follows:

$$\int_0^\infty f(x) dx = 1 . \tag{5.6}$$

This reflects the fact that one electron is excluded from the Fermi hole. For the gas of non-interacting electrons, $f(x)$ is known exactly:

$$f(x) = 9(\sin x - x \cos x)^2 / x^6 . \tag{5.7}$$

Using (5.4) this leads to the exchange energy

$$E_x = -3\left(\frac{3}{4\pi}\right)^{1/3} [\rho_\uparrow^{4/3} + \rho_\downarrow^{4/3}], \tag{5.8}$$

and hence to the exchange potential

$$V_x^\sigma(\boldsymbol{r}) = -\frac{2}{3} \times 6 \left(\frac{3}{4\pi} \rho_\sigma(\boldsymbol{r})\right)^{1/3}, \tag{5.9}$$

which is known as the *Gaspar–Kohn–Sham* potential [5.5, 6]. When $f(x)$ is different from (5.7), the potential V_x^σ differs from (5.9) only by a constant factor. Instead of the factor of 2/3 in (5.9), *Slater* proposed the adjustable parameter α, which could be varied from 2/3 to 1. This last value corresponds to high density of the electron gas. The formula of the exchange-correlation potential is then

$$V_x^\sigma(\boldsymbol{r}) = -6\alpha \left[\frac{3}{4\pi} \rho_\sigma(\boldsymbol{r})\right]^{1/3} \tag{5.10}$$

and constitutes the essence of the well-known X_α method [5.7].

The reasonable choice of α values for various calculations was discussed elsewhere [5.8] when the X_α method became popular. Apart from its simplicity as compared with the Hartree–Fock method (due to the fact that the exchange potential is now local), there is some evidence that the potential (5.10) actually includes the correlation effects and therefore somewhow goes beyond the Hartree-Fock approximation. This judgement was initially based mostly on empirical consderations, but later on this subject became clearer in a series of detailed studies of correlation effects.

While the exchange potential for the chosen spin direction depends on the charge density of this spin only, the correlation part of V_{xc} is, in principle, influenced by both spin densities, because the interaction of the electron with the "correlation hole" has an essentially many-particle origin.

The exchange energy per particle $\varepsilon_x(\rho_\uparrow, \rho_\downarrow)$ which leads to (5.8) can be found as an interpolation between two limiting values corresponding to the cases of "paramagnetic" ($\rho_\uparrow = \rho_\downarrow$) and "ferromagnetic" ($\rho_\uparrow = 0$) electron gas:

$$\varepsilon_x(r_s, \zeta) = \varepsilon_x(r_s, 0) + [\varepsilon_x(r_s, 1) - \varepsilon_x(r_s, 0)] f(\zeta), \tag{5.11}$$

where

$$\zeta = (\rho_\uparrow - \rho_\downarrow)/\rho, \rho = \rho_\uparrow + \rho_\downarrow, r_s = (4\pi\rho/3)^{-1/3},$$

$$\varepsilon_x(r_s, 0) = -\frac{3}{\pi\left(\frac{4}{9\pi}\right)^{1/3} r_s},$$

$$\varepsilon_x(r_s, 1) = -\frac{3}{\pi\left(\frac{2}{9\pi}\right)^{1/3} r_s}, \tag{5.12}$$

with the following interpolation formula:

$$f(\zeta) = \frac{(1+\zeta)^{4/3} + (1-\zeta)^{4/3} - 2}{2(2^{1/3} - 1)}. \tag{5.13}$$

Von Barth and *Hedin* [5.9] proposed using the same interpolation formula for the correlation energy $\varepsilon_c(r_s, \zeta)$ as well. The problem is reduced then to the evaluation of ε_c for "paramagnetic" ($\zeta = 0$) and "ferromagnetic" ($\zeta = 1$) Fermi liquid. Based on the type of interpolating function introduced by *Hedin* and *Lundqvist* [5.10], different parametrizations have been proposed [5.9, 11, 12]:

$$\varepsilon_c(r_s) = -C_i\left[(1 + x_i)^3 \ln\left(1 + \frac{1}{x_i}\right) + \frac{x_i}{2} - x_i^2 - \frac{1}{3}\right]. \tag{5.14}$$

Here i corresponds to the paramagnetic or ferromagnetic case, $x_i = r_s/r_i$, and C_i and r_i are the parameters chosen to fit some estimates of $\varepsilon_c(r_s, 0)$ and $\varepsilon_c(r_s, 1)$. Corresponding estimations of these parameters have been made [5.9] in the random-phase approximation. *Gunnarsson* and *Lundqvist* [5.11] used another approximation, with correlations treated in terms of the linear-response approach. After *Ceperley* and *Alder* [5.13] performed Monte Carlo calculations of ground-state energies for many-electron systems at different densities, several attempts were undertaken to find the best analytical interpolation which would fit their "exact" results. A prescription by *Vosko* et al. [5.14], which is one of the most widely used up to now, is based on a Padé approximation and defines the one-particle correlation energy as follows:

$$\varepsilon_c(r_s) = A\left\{\ln\frac{x^2}{X(x)} + \frac{2b}{Q}\arctan\frac{Q}{2x+b} - \frac{bx_0}{X(x_0)}\left[\ln\frac{(x-x_0)^2}{X(x)} + \frac{2(b+2x_0)}{Q}\arctan\frac{Q}{2x+b}\right]\right\}. \tag{5.15}$$

Here $x \equiv r_s^{1/2}$, $X(x) \equiv x^2 + bx + c$, $Q \equiv (4c - b^2)^{1/2}$, and the parameters are $A = 0.0621814$ Ry, $x_0 = -0.409286$, $b = 13.0720$, $c = 42.7198$ for the paramagnetic case; and $A = 0.0310907$, $x_0 = -0.743294$, $b = 20.1231$, $c = 101.578$ for the ferromagnetic case.

For the dependence of ε_c on the spin polarization in the intermediate region from $\zeta = 0$ to $\zeta = 1$, *Vosko* et al. [5.14] proposed several better choices than the one arising from the von Barth–Hedin prescription (5.11). For practical calculations, they recommended

$$\varepsilon_c(r_s, \zeta) = \varepsilon_c(r_s, 0) + \alpha_c(r_s)[f(\zeta)/f''(0)][1 + \beta(r_s)\zeta^4], \tag{5.16}$$

where

$$\beta(r_s) = \frac{f''(0)[\varepsilon_c(r_s, 1) - \varepsilon_c(r_s, 0)]}{\alpha_c(r_s)} - 1. \tag{5.17}$$

The spin stiffness constant $\alpha_c(r_s)$ is defined by the same interpolation formula (5.15) as was used for $\varepsilon_c(r_s)$, but with other coefficients: $A = -1/(3\pi^2)$ Ry, $x_0 = -0.00475840$, $b = 1.13107$, $c = 13.0045$.

When defined in one or another way, the one-particle correlation energy $\varepsilon_c(r_s, \zeta)$ together with the exchange energy $\varepsilon_x(r_s, \zeta)$ determines the exchange-correlation energy functional – in the local approximation (5.4). Hence the local exchange-correlation potential is

$$V_{xc}^{\sigma}(\boldsymbol{r}) = \frac{\delta E_{xc}[\rho_{\uparrow}(\boldsymbol{r}), \rho_{\downarrow}(\boldsymbol{r})]}{\delta\rho_{\sigma}(\boldsymbol{r})}. \tag{5.18}$$

The major part of its dependence on $\rho_\sigma(\boldsymbol{r})$ can be described as $\sim \rho_\sigma(\boldsymbol{r})^{1/3}$; i.e. it is already included in the X_α approximation (5.10). This is why the latter appeared to be so successful. All subsequent sophisticated expressions for the exchange-correlation potential helped to specify the relatively minor variation of the coefficient in (5.10) with both spin-up and spin-down densities. This smooth dependence reveals the effect of the shape of the "Fermi hole" and "correlation hole".

The local approximation for the exchange-correlation energy (5.4), although quite satisfactory for many applications, bears a source of unpredictable error in itself. Two major difficulties are the bad treatment of non-uniform charge distributions, and the non-sufficient cancellation of the Coulomb self-interaction of an electron. Both problems are closely related.

The first was long thought to be unimportant for the charge densities typical for metal systems. However, several evidently wrong results obtained in band-structure calculations (such as the conclusion that the fcc phase of iron has lower total energy than the bcc phase) were proved recently to be the drawbacks of the local approximation used [5.15]. To overcome these difficulties, several prescriptions for non-local exchange-correlation energy functionals have been proposed, e.g. those by *Langreth*, *Mehl* and *Hu* [5.16–18] or by *Perdew* and *Wang* [5.19–21]. They were tested in practical calculations [5.15].

The problem of overestimating the Coulomb self-interaction appears in the LSDA formalism as the result of averaging over individual orbitals when switching to the spin density as the basic variable. Actually, the spherically symmetric Fermi hole is too crude to imitate the spin density that must be excluded from the interaction with the given electron. This problem does not arise in the Hartree-Fock approximation with its exact orbital-dependent exchange potential. Any attempt to improve the situation in the spin-density functional approach demands also the inclusion of orbital-dependent correction to the local exchange-correlation function, as proposed by *Perdew* and *Zunger* [5.22]:

$$E_{xc}[\rho_\uparrow, \rho_\downarrow] = E_{xc}^{\text{approx}}[\rho_\uparrow, \rho_\downarrow] - \sum_{i\sigma} \delta_{i\sigma}.$$

$$\delta_{i\sigma} = \int d\boldsymbol{r}\, d\boldsymbol{r}' \frac{\rho_{i\sigma}(\boldsymbol{r})\rho_{i\sigma}(\boldsymbol{r}')}{|\boldsymbol{r} - \boldsymbol{r}'|} + E_{xc}^{\text{approx}}[\rho_{i\sigma}, 0], \tag{5.19}$$

where $E_{xc}^{\text{approx}}[\rho_\uparrow, \rho_\downarrow]$ is any approximation to the unknown exchange-correlation functional (for example, the energy defined in the local approximation), and $\delta_{i\sigma}$ is the self-interaction correction (SIC), which ensures the cancellation of Coulomb self-interaction for the orbital i with the spin σ and spin density $\rho_{i\sigma}(\boldsymbol{r})$.

The upper estimate of the self-interaction correction found by *Perdew* and *Zunger* [5.22] is as follows:

$$\delta_{i\sigma} \leqslant 0.16 \int d\boldsymbol{r}\, \rho_{i\sigma}^{4/3}(\boldsymbol{r}). \tag{5.20}$$

As is seen, SIC becomes more important for localized states. Also, SIC is non-zero only for occupied states and lowers their energy. Its inclusion leads to especially pronounced effects in semiconductors and insulators where empty and filled states are quite clearly separated in the energy spectrum. Actually, the LSDA-based calculations for such systems usually fail to reproduce the correct (experimentally measured) value of the band gap (in some cases, even the existence of the band gap) and hence underestimate the tendency for localization and the values of local magnetic moments. Then, the inclusion of SIC is vital to help the situation.

Many of the ab initio calculations performed up to now have used a muffin-tin or another similar approximation for the potential, assuming a spherically averaged form of the potential within atomic cells (see the next section). The question therefore arises, whether non-local energy functionals as a whole, or the self-interaction corrected functional as an example, can be implemented within these computational schemes. Preliminary results available show that the non-local functionals, even if they are non-symmetric by definition, retain much of their usefulness in applications which deal with the spherically averaged potential. Thus, in the case of SIC, the possibility for different orbitals to feel different potentials is more important than precision in treating the spatial shape of these potentials. The implementation of SIC in any simple form helps to overcome most drawbacks of the LSDA. However, the inclusion of SIC even in the simplest form leads to severe computational problems, like non-orthogonality of the orbitals, which are defined now by actually different Hamiltonians [5.23].

5.2 Band-Structure Approaches in the Green Function Formalism

A variety of the most practically important physical properties of crystals can be described in terms of their one-electron Green functions. For one-electron density of states $n_\sigma(E)$ and spatial charge density $\rho_\sigma(\boldsymbol{r})$, the following expressions are well known:

$$n_\sigma(E) = -\frac{1}{\pi} \int_{\Omega_{WS}} \mathrm{Im}\{G_\sigma(\boldsymbol{r}, \boldsymbol{r}, E)\}\, d\boldsymbol{r}, \tag{5.21}$$

$$\rho_\sigma(\boldsymbol{r}) = -\frac{1}{\pi} \int_{\infty}^{E_F} \mathrm{Im}\{G_\sigma(\boldsymbol{r}, \boldsymbol{r}, E)\}\, dE. \tag{5.22}$$

Calculations of other measurable properties are discussed in the next section.

The Green function is related to the crystal potential via the Lippmann-Schwinger equation:

$$G(\boldsymbol{r}, \boldsymbol{r}', E) = g(\boldsymbol{r}, \boldsymbol{r}', E) + \int d\boldsymbol{r}''\, g(\boldsymbol{r}, \boldsymbol{r}'', E)\, V(\boldsymbol{r}'')\, G(\boldsymbol{r}'', \boldsymbol{r}', E), \tag{5.23}$$

where $g(\boldsymbol{r}, \boldsymbol{r}', E)$ is the Green function of free electrons:

$$g(\boldsymbol{r}, \boldsymbol{r}', E) = -\frac{1}{4\pi} \frac{e^{i\sqrt{E}|\boldsymbol{r}-\boldsymbol{r}'|}}{|\boldsymbol{r}-\boldsymbol{r}'|}. \tag{5.24}$$

The straightforward solution of (5.23) appears to be extremely difficult. The problem, however, becomes feasible when the muffin-tin approximation, or the space-filling atomic-spheres approximation, is used for the potential. These approximations are considered to be well justified for closely packed structures; however, the addition of "empty spheres" with spherically symmetric potential in interstitial regions allows one to use this approximation in many cases of loosely packed structures as well.

If the potential $V_i(\boldsymbol{r})$ within the spheres centred at $\boldsymbol{R}_i$ is spherically symmetric, the Green function (5.23) can be expanded over the basis functions $Y_L(\boldsymbol{r} - \boldsymbol{R}_i)$ $R_l^i(|\boldsymbol{r} - \boldsymbol{R}_i|, E)$, which are the solutions of the Schrödinger equation inside the spheres. Namely, $R_l^i(r, E)$ is supposed to be a regular solution of the radial Schrödinger equation with the following asymptotic behaviour:

$$R_l^i(r, E)_{r \geq R_{MT}} \to \cos\delta_l^i(E) j_l(\sqrt{E}r) - \sin\delta_l^i(E) n_l(\sqrt{E}r). \tag{5.25}$$

Then, the problem reduces to the evaluation of energy-dependent matrix elements of the Green function $G_{LL'}^{ij}(E)$ using the Dyson equation:

$$G_{LL'}^{ij}(E) = g_{LL'}^{ij}(E) + \sum_{kL_1} g_{LL_1}^{ik}(E)\, t_{l_1}^k(E)\, G_{L_1L'}^{kj}(E). \tag{5.26}$$

Summation over the sites k includes, in principle, the sites of the whole crystal. In the summation over orbital indices $L \equiv (l, m)$, in most cases only the terms up to $l = 3$ must be retained, because these sums are known to converge rapidly.

The one-centre scattering matrices $t_l^i(E)$ can be evaluated easily, once the l-dependent phase shift of the wave of given energy $\delta_l^i(E)$ is found from the condition that the radial solution $R_l^i(r, E)$ matches the asymptotic tail (5.25) smoothly at the muffin-tin (or atomic-sphere) surface:

$$t_l^i(E) = -\frac{1}{\kappa}\sin\delta_l^i \exp(i\delta_l^i). \tag{5.27}$$

Matrix elements of the free-electron Green function (5.24) are known to be

$$g_{LL'}^{ij}(E) = -i\sqrt{E}\sum_{L_1} I_{L_1}(L, L')\cdot 4\pi i^{l_1} h_{l_1}(\sqrt{E}|\boldsymbol{R}_i - \boldsymbol{R}_j|)\, Y_{L_1}(\boldsymbol{R}_i - \boldsymbol{R}_j). \tag{5.28}$$

Here $I_{L_1}(L, L') = \int d\hat{\boldsymbol{r}}\, Y_{L_1}(\hat{\boldsymbol{r}})\, Y_L(\hat{\boldsymbol{r}})\, Y_{L'}(\hat{\boldsymbol{r}})$ are Gaunt coefficients (real spherical harmonics are used hereafter), and h_l are spherical Hankel functions. Once the matrix elements of the Green function are found by solving (5.26), its total spatial dependence can be described as follows:

$$G(\boldsymbol{R}_i + \boldsymbol{r}, \boldsymbol{R}_j + \boldsymbol{r}', E) = \sum_{LL'} R_l^i(r, E)\, Y_L(\boldsymbol{r}) e^{i\delta_l^i(E)} G_{LL'}^{ij}(E) e^{i\delta_l^j(E)} R_{l'}^j(r', E)\, Y_{L'}(\boldsymbol{r}')$$

$$+\, \delta_{ij}\sqrt{E}\sum_L R_l^i(r_<, E)\, Y_L(\boldsymbol{r}) H_l^j(r_>, E)\, Y_L(\boldsymbol{r}'). \tag{5.29}$$

Here $r_>$ and $r_<$ are the minimum and the maximum of two distances (r, r'). The complex function $H_j^i(r, E)$, which is a combination of the regular and non-regular solutions of the Schrödinger equation, is defined in [5.24].

The energy in (5.21–29) is supposed to be positive and measured with respect to the "muffin-tin zero", or the level of interstitial constant potential (for the muffin-tin approximation) that separates the discrete spectrum from the continuum. In the atomic-spheres approximation, there is no interstitial space, and it is convenient to set $\kappa \equiv \sqrt{E} = 0$. The corresponding changes in the formulas are discussed below.

We discuss first the possible ways of solving (5.26). The crudest way is to restrict the summation over the sites of a finite cluster of atoms. The rest of the crystal is neglected or taken into account approximately, via its effect on the potential inside the spheres chosen, or via the appropriate boundary conditions. The problem, then, is solved by inverting the matrix of a finite size:

$$\boldsymbol{G} = [\boldsymbol{g}^{-1} - \boldsymbol{t}]^{-1}. \tag{5.30}$$

To simplify the calculations and avoid performing matrix inversion twice, one can introduce the scattering path operator

$$\boldsymbol{T} = [\boldsymbol{t}^{-1} - \boldsymbol{g}]^{-1}, \tag{5.31}$$

which defines the Green function as follows:

$$G^{ij}_{LL'}(E) = g^{ij}_{LL'}(E) + \sum_{\substack{k \neq i \\ k' \neq j}} \sum_{\substack{L_1 \\ L_2}} g^{ik}_{LL_1}(E) T^{kk'}_{L_1L_2}(E) g^{k'j}_{L_2L'}(E). \tag{5.32}$$

For practical calculations, (5.31) is more attractive than (5.30) because the inner matrix to be inverted (t) is diagonal.

The cluster Green function method [5.25, 26] gained a certain popularity due to its relative simplicity. There is, however, a problem that can hardly be eliminated in this approach, namely, the very poor convergence of results as the cluster size increases. Even a cluster of several tens of atoms fails to reproduce the detailed structure of the density of states of an infinite perfect crystal and its related properties. The situation is, however, not as bad if only integrated properties, such as magnetic moments, are studied, or if the total shape of the DOS, rather than its fine details, is of interest. The latter relates to the interpretation of X-ray or photoelectron spectra, which are already experimentally broadened, or to the study of the electronic structure of disordered alloys. These two cases mostly cover the areas where the cluster approach is applied successfully.

An alternative approach is based on the lattice Fourier transformation which replaces the real-space summation by integration over the Brillouin zone Ω_{BZ}:

$$G^{ij}_{LL'}(E) = \frac{1}{\Omega_{BZ}} \int d\boldsymbol{k}\, G_{LL'}(\boldsymbol{k}, E) \exp[i\boldsymbol{k}(\boldsymbol{R}_i - \boldsymbol{R}_j)]. \tag{5.33}$$

The Green function elements can be found by inverting the finite-size matrix

$$G_{LL'}(\boldsymbol{k}, E) = \sum_{L_1} [1 - g(\boldsymbol{k}, E) t(E)]^{-1}_{LL_1} g_{L_1L'}(\boldsymbol{k}, E). \tag{5.34}$$

The poles of the Green function define the energy spectrum for a given $\boldsymbol{k}$ point:

$$|t^{-1}(E) - g(\boldsymbol{k}, E)| = 0. \tag{5.35}$$

This is the well-known secular equation of the *Korringa-Kohn-Rostoker* method (KKR) [5.27, 28]. Here the coefficients

$$g_{LL'}(\boldsymbol{k}, E) = \sum_i g^{0i}_{LL'}(E) \exp[i\boldsymbol{k} \cdot \boldsymbol{R}_i] \tag{5.36}$$

do not depend on the potential; they are the structure constants of the KKR method.

Although such a scheme requires that a crystal possess translational symmetry, the combination of it with the above-mentioned cluster or real-space approach gives an extremely powerful calculation method applicable to almost all cases of crystals with point defects. (By "point defects" we mean here not only substitutional impurities or vacancies, but also local magnetic fluctuations, like sites having local magnetic moments that are inverted or deviate from the main direction).

The essence of this unified method – known as the KKR–Green function method [5.29] – lies in the fact that the Lippmann–Schwinger equation (5.23) is still valid if $g(\boldsymbol{r}, \boldsymbol{r}', E)$ is replaced by the Green function of perfect crystal, and $V(\boldsymbol{r})$ by the deviation of the potential from that of a perfect crystal. Then $G(\boldsymbol{r}, \boldsymbol{r}', E)$ gives the Green function of a defect crystal. In terms of (5.30) this means that $\Delta t = \tilde{t}_l^i(E) - t_l^i(E)$ is to be substituted for $\tilde{t}$, indicating the difference of the scattering matrix elements from their values for the case of a perfect crystal. These differences are non-zero only for a finite number of defect sites; therefore, once the Green function of the perfect crystal (5.34) is found, the Green function for the defect crystal can be calculated explicitly by inversion of the finite-size matrix (5.30). When solving an actual computational problem for a given defect system, such a last step (recalculation of the defect-crystal Green function) has to be performed iteratively, until self-consistency is achieved [5.29]. This process is no more time consuming than the usual cluster calculation, because the Green function of the perfect crystal is calculated in advance, once and for all. The evaluation of this Green function is, however, rather troublesome due to the necessity to invert a complex matrix for many E and $\boldsymbol{k}$ values.

Such a procedure becomes easier in the LMTO-ASA formalism. So far, we made no distinction between the cases involving muffin-tin and atomic-sphere types of cells. We mentioned only that the limit $\kappa \to 0$ taken in the atomic-spheres approximation changed some of the formulas involved. The transition to the formalism of the atomic-spheres approximation (ASA) can be performed starting from the KKR equations, as described by *Andersen* [5.30] and *Koenig* et al. [5.31]. As a result, one arrives at the following KKR-ASA equations:

$$[p(E) - S(\boldsymbol{k})][\dot{p}^{-1/2}(E)]B_{\boldsymbol{k}} = 0, \tag{5.37}$$

where $S(\boldsymbol{k})$ are energy-independent structure constants in the ASA method, and $p(E)$ is the $\kappa \to 0$ limit of the inverse t matrix (5.27), which is called the potential function. Actually, a transition from the KKR secular equation to (5.37) supposes the bracketing of all involved matrices by $A_l^{-1}(\kappa) \ldots A_{l'}^{-1}(\kappa)$, where

$$A_l = \frac{(2l-1)!!}{\sqrt{2S}(\kappa S)^l} \tag{5.38}$$

(S is the atomic-sphere radius), before taking the limit $\kappa \to 0$. The KKR eigenstates are transformed in a similar way by multiplying them by (5.38) once and taking the $\kappa \to 0$ limit. This procedure results in an additional factor $[\dot{p}^{1/2}(E)]$ (the dot denotes the energy derivative); see (5.37). This has been discussed in more detail elsewhere [5.32, 33].

The energy spectrum is determined then by the secular equation

$$|p(E) - S(\boldsymbol{k})| = 0. \tag{5.39}$$

The potential function $p(E)$ depends on the potential inside the given atomic sphere only, and it is usually expressed in terms of $D_l(E)$ – logarithmic derivatives of $R_l^i(r, E)$ at the atomic-sphere boundary S:

$$p_l(E) = 2(2l+1)\frac{D_l(E)+l+1}{D_l(E)-l}. \tag{5.40}$$

Of course, the energy independence of structure constants within ASA makes the practical calculations easier. In fact, the ASA structure constants do not depend on the actual lattice spacing either. The latter fact leads to interesting consequences, like the possibility of plotting and analysing canonical energy bands and densities of states. They reveal features relevant to the given type of crystal structure, with the effect of the potential and hybridization separated from them.

Even more important results concerning the feasibility of the scheme proposed to calculate the Green function follow from the linearized version of the band-structure equations. This is based on the physically well-grounded idea proposed by *Andersen* [5.30]. The radial wavefunction within the atomic spheres $R_l^i(r, E)$ varies with energy very smoothly. Then, if one expands it into a Taylor series (around an arbitrarily chosen energy, usually taken near the centre of the corresponding energy band) and keeps only zero-order terms and terms linear in energy, the error introduced is of the order of 10^{-2} of the magnitude of the wavefunction [5.30].

Important properties resulting from this concern the potential function (5.40), which can be parametrized, with only several parameters being introduced to simulate its overall behaviour in a broad energy range. When $D_l(E)$ is expressed in terms of the radial solution and its energy derivative, both taken at fixed energy E_ν, one can obtain easily

$$[D(E) - D_\nu]^{-1} = -\frac{1}{mS^2\varepsilon} + a + bS^2\varepsilon. \tag{5.41}$$

Here D_ν is the logarithmic derivative of the radial wave function at the energy E_ν; $\varepsilon \equiv E - E_\nu$; S is the atomic-sphere radius; and a, b, and m are the potential parameters (different for different l values), which are combinations of $R(S, E)$, $\partial R(S, E)/\partial E$ at $E = E_\nu$, and S. The last term in (5.41) can usually be neglected. This leads to the following simplified expression for $p_l(E)$:

$$p_l(E) = \frac{\Gamma}{V - E} + Q, \tag{5.42}$$

where Γ, V, and Q are other potential parameters. Classification of all these potential parameters and the details of derivation of corresponding formulas can be found in *Skriver*'s book [5.34].

If the potential function is taken to have a special parametrized form (5.42), it becomes possible to introduce explicitly the Hamiltonian of the system under consideration [5.35]. Substitution of (5.42) into (5.37) results in

$$[p - \boldsymbol{S}]\dot{p}^{-1/2} = (Q - \boldsymbol{S})[\varepsilon - V - \Gamma^{1/2}(Q - \boldsymbol{S})^{-1}\Gamma^{1/2}]\Gamma^{-1/2}, \tag{5.43}$$

where p, Q, V, and Γ are matrices diagonal over the sites and l. As is seen, the combination

$$\boldsymbol{H} = V + \Gamma^{1/2}(Q - \boldsymbol{S})^{-1}\Gamma^{1/2} \tag{5.44}$$

plays the role of a Hamiltonian, because it allows us to present (5.37) in the form of the usual Schrödinger equation:

$$(E - \boldsymbol{H})\boldsymbol{B} = 0. \tag{5.45}$$

The possibility of working with the explicitly determined Hamiltonian is a pleasant feature of tight-binding-type calculation schemes (but it is not available in the KKR and other related multiple-scattering schemes). This is not especially important if one is interested in solving the band-structure problem only, but simplifies considerably the Green function calculation. It is no longer necessary to perform the matrix inversion in (5.34) for many energy values. Instead, the real-space Green function can be constructed, making use of its spectral representation:

$$G(\boldsymbol{r}, \boldsymbol{r}', E) = \frac{1}{\Omega_{\mathrm{BZ}}} \int d\boldsymbol{k} \sum_n \frac{\Psi_{n\boldsymbol{k}}(\boldsymbol{r})\Psi^*_{n\boldsymbol{k}}(\boldsymbol{r}')}{E - E_n(\boldsymbol{k}) + i0}. \tag{5.46}$$

Actually, we are interested in calculation of the site-numbered matrix elements, which are expressed in terms of eigenvectors:

$$G^{(it)(jt')}_{LL'}(z) = \frac{1}{\Omega_{\mathrm{BZ}}} \int d\boldsymbol{k} \sum_n \frac{B^t_{nL}(\boldsymbol{k}) B^{t'}_{nL'}(\boldsymbol{k}) \exp[i\boldsymbol{k}\cdot(\boldsymbol{R}_i - \boldsymbol{R}_j)]}{z - E_n(\boldsymbol{k})} \tag{5.47}$$

Here the matrix element of the Green function is written between the function with given $L \equiv (l, m)$, centred at atom t in the cell i, and the function with L' at atom t' in the cell j; $\boldsymbol{R}_i$ and $\boldsymbol{R}_j$ are lattice vectors. To show that the energy variable can be complex, z is written instead of E.

It is possible to restrict $\boldsymbol{k}$-space integration in (5.47) to the irreducible wedge of the Brillouin zone only, including the summation over the symmetry operations of the corresponding crystal space group in the nominator. The transformations of the eigenvectors which should be rotated can be expressed in terms of the Wigner matrices [5.36].

An alternative algorithm for calculating the Green function (applicable when one is interested in the values of the latter only along the real energy axis only) consists of finding first the imaginary part of (5.47):

$$\begin{aligned}\mathrm{Im}\{G^{(it)(jt')}_{LL'}(E)\} = &-\frac{\pi}{\Omega_{\mathrm{BZ}}} \int d\boldsymbol{k} \sum_n \mathrm{Re}\{B^t_{nL}(\boldsymbol{k}) B^{t'}_{nL}(\boldsymbol{k}) \exp[i\boldsymbol{k}(\boldsymbol{R}_i - \boldsymbol{R}_j)]\} \\ &\times \delta[E - E_n(\boldsymbol{k})].\end{aligned} \tag{5.48}$$

The real part is restored afterwards via the Hilbert transformation:

$$\mathrm{Re}\{G^{(it)(jt')}_{LL'}(E)\} = -\mathscr{P} \int \frac{dE'}{E - E'} \mathrm{Im}\{G^{(it)(jt')}_{LL'}(E')\}. \tag{5.49}$$

Making use of the spectral representation, it is no longer necessary to determine the Green function via the scattering path operator according to (5.31, 32). However, the scattering path operator itself is very important for some applications, as discussed in the next section. In the LMTO-ASA approach with the potential function parametrized as in (5.42), it is possible to find a simple expression for $\boldsymbol{T} = (p - \boldsymbol{S})^{-1}$ [5.35]:

$$T_{LL'}^{(it)(jt')}(E) = \frac{E - V_i}{\Gamma_i}\delta_{(it)(jt')} - \frac{E - V_i}{\Gamma_i^{1/2}} G_{LL'}^{(it)(jt')}(E) \frac{E - V_j}{\Gamma_j^{1/2}}. \tag{5.50}$$

After this brief discussion of the advantages of the LMTO-ASA approach in the calculation of the perfect-crystal Green function, let us turn to the problem of finding the property for a defect crystal. As we mentioned above, the approach is based on the Dyson equation (5.26), or its other form (5.32), which we now write as

$$\tilde{G}_{LL'}^{ij}(E) = G_{LL'}^{ij}(E) + \sum_{\substack{k \neq i \\ k' \neq j}} \sum_{\substack{L_1 \\ L_2}} G_{LL_1}^{ik}(E)(\Delta T)_{L_1 L_2}^{kk'}(E) G_{L_2 L'}^{k'j}(E). \tag{5.51}$$

Here $\tilde{G}$ is the Green function of the crystal with defects, and (ΔT) is the "residual" scattering path operator which takes into account the scattering of electrons at the distortions of the perfect-crystal potential:

$$(\Delta \boldsymbol{T}) = [(\Delta t)^{-1} - \boldsymbol{G}]^{-1}. \tag{5.52}$$

This formula resembles (5.31), but (Δt) now denotes the difference between individual t matrices (or, correspondingly, the reversed potential functions in the atomic-spheres approximation) of defect and perfect crystals.

The summation over sites in (5.51) is restricted to a finite region of, say, N sites, which includes the perturbation of the potential due to the presence of the defects, i.e. the "embedded cluster". Here (ΔT) is evaluated by the inversion of an $N(l_{\max} + 1)^2$ size matrix, whee $l_{\max}$ denotes the maximal l value retained in the expansions over partial waves. Therefore, the solution of the Dyson equation is computationally no more difficult than the usual cluster calculation. In fact, since the information on the long-range order in the crystal is already included in the "perfect-crystal" Green function, there is no need to include in the cluster a large number of atoms merely to imitate an infinite crystal. The region of really disturbed potential around the point defect usually does not extend further than second neighbours (the magnetic impurities in Pd are a notable exception); therefore, the convergence of the results over the cluster size is usually rather fast. For some systems, reasonable results can be obtained with only one atom included in the cluster, i.e. the defect itself (Chap. 6).

The method of solving the Dyson equation is, in principle, the same in the KKR or LMTO-ASA formalisms. In the latter approach, where a parametrized form for the potential function (5.42) is used to formulate explicitly and solve the eigenstates problem (5.45), it is no longer necessary to retain the same form when coming to the Dyson equation. Instead, another more accurate parametrization or the exact expression for the potential function (5.40) can be used.

Equation (5.51) has to be solved iteratively, until the potentials at defect sites converge. Computational effort in the iteration process can be saved in two ways. First, the symmetry properties of the embedded cluster can be used, enabling one to transform the matrix to be inverted in (5.52) to block-diagonal form. Second, the energy integration of the Green function – as in (5.22), to obtain the charge density and potential for the next iteration – may be performed along the complex energy path, where the function behaves more smoothly. Both possibilities have been implemented and described [5.29, 37].

5.3 Magnetic Interactions Within the LSDA

As has been mentioned above, many important physical parameters can be defined in terms of the matrix elements of the Green function. We briefly describe below a possible way to derive corresponding relations. As an example the magnetic interaction parameters will be considered, although the method is quite general and can be applied to many other characteristics of solids.

The magnetic interactions we would like to consider give rise to some changes in the total energy of the system. Physical models establish the connection between the total energy variations and the external variables within the model under consideration, such as individual spins, angles and distances between them. The connections are formulated in terms of model parameters. When calculated from first principles, they would help much in understanding the underlying process, discussion of the experimental results, etc. The idea of the calculation is based on the fact that corresponding changes in the total energy, or the derivative of the latter with respect to different external variables, can often be calculated independently from first principles.

An attempt to calculate the corresponding energy differences explicitly usually fails to provide sufficient precision, since the total energies which have to be compared are several orders of magnitude larrger than purely magnetic contributions to them. A possible way to overcome this difficulty is to treat the corresponding variations of the total energy analytically on the basis of the so-called local force theorem. In this approach, one is restricted to small fluctuations from the ground state, but this approach is quite sufficient derivation of most of the physically important parameters.

We proceed from the most general expression for the total energy:

$$E_{tot} = \sum_{i} \varepsilon_i - E_{dc} , \tag{5.53}$$

where the first term denotes one-electron energies summed up over the occupied states, and the second term cancels the interaction energies doubly counted in the first sum:

$$E_{dc} = \int dr\, dr' \frac{\rho(\boldsymbol{r})\rho(\boldsymbol{r}')}{|\boldsymbol{r} - \boldsymbol{r}'|} + \int dr\, \rho^2(\boldsymbol{r}) \frac{d\varepsilon_{xc}}{d\rho}. \tag{5.54}$$

The total energy variation can be expressed as a sum of the following terms: $\delta^* E_{sp}$, i.e. the variation of the single-particle energies with the electron density kept fixed; $\delta^{pot} E_{sp}$, i.e. the variation due to changes of the potential; and $-\delta E_{dc}$. It has been shown [5.38–40] that the two latter contributions cancel almost completely, and the remaining term is of second order in the excitation considered:

$$\delta E = \delta^* E_{sp} - \int \delta \boldsymbol{S}\, d\boldsymbol{S}\, \rho^2(\boldsymbol{r}) \frac{d\varepsilon_{xc}}{d\rho}. \tag{5.55}$$

The second term, i.e. the error of the cancellation, takes into account the change in the exchange-correlation energy due to the variation of volume. It can be neglected in the applications discussed below.

Therefore, the variation of the total energy can be expressed through the variations of one-particle energies, provided they are calculated with the non-perturbed charge density of the ground state. Equation (5.55) can be written as

$$\delta E_{tot} = \int^{E_F} E\, \delta n(E)\, dE = -\int^{E_F} \delta N(E)\, dE, \tag{5.56}$$

where the number-of-states function $N(E) = \int^E n(E')\, dE'$ is introduced. Here $N(E)$ is related to the scattering path operator $\boldsymbol{T}$ by *Lloyd*'s formula [5.41]:

$$N(E) = \frac{1}{\pi} \operatorname{Im}\{\operatorname{Tr}\{\ln \boldsymbol{T}(E)\}\}. \tag{5.57}$$

If $\boldsymbol{T}$ denotes the scattering path operator matrix in the ground state, then in the presence of a local excitation it becomes

$$\tilde{\boldsymbol{T}} = \boldsymbol{T}[1 + \delta\boldsymbol{\tau}\boldsymbol{T}]^{-1}, \tag{5.58}$$

where $\delta\tau = \tilde{t}^{-1} - t^{-1}$, i.e. the change of the reverse t matrix on one or more sites due to the excitation. As a result, one gets the following expression for the variation of the total energy:

$$\delta E_{tot} = \frac{1}{\pi} \int^{E_F} \operatorname{Im}\{\operatorname{Tr}\{\ln[1 + \delta\boldsymbol{\tau}\boldsymbol{T}]\}\}\, dE. \tag{5.59}$$

This general formula can be used to derive expressions for various physical parameters, provided the type of excitation is specified in terms of $\delta\boldsymbol{\tau}$.

We consider first the excitation from the ground state of a ferromagnetic (or an antiferromagnetic) system, namely, a deviation of the local magnetic moment of an atom from its initial direction collinear to all other moments. This situation can be analysed within the *Heisenberg* model, with its well-known expression for the exchange energy [5.42]:

$$E_{ex} = \sum_{i \neq j} J_{ij} \boldsymbol{e}_i \boldsymbol{e}_j. \tag{5.60}$$

The variation of the exchange energy due to the deviation of the local moment at the site $i = 0$ by the angle θ is

$$\delta E_{ex} = 2 \sum_j J_{0j}(1 - \cos\theta) \simeq J_0 \theta^2, \tag{5.61}$$

where J_0 is the effective exchange parameter which defines the total interaction of an atom with the ferromagnetic (or antiferromagnetic) background. This is closely related to the Curie (or Neel) temperature in simplest case of the mean-field approximation by

$$T_C = \frac{2}{3k} J_0 \eta. \tag{5.62}$$

Here η is the so-called quantum factor, which equals $(S + 1)/S$ (S being the spin value) in the Heisenberg model. It must be set equal to 1 in the classical limit.

To calculate J_0 from first principles, one needs to express $\delta\tau$ in terms of local deviations θ. Rotations in the spin space can be most easily treated using decompositions of $\delta\boldsymbol{\tau}$ into the Pauli matrices. Thus,

$$\boldsymbol{\tau} = \tfrac{1}{2}(\boldsymbol{\tau}_\uparrow + \boldsymbol{\tau}_\downarrow) \times \boldsymbol{1} + \tfrac{1}{2}(\boldsymbol{\tau}_\uparrow - \boldsymbol{\tau}_\downarrow) \times (\boldsymbol{e} \cdot \boldsymbol{\sigma}), \tag{5.63}$$

where $\tau_\uparrow$ and $\tau_\downarrow$ are spin-up and spin-down inverse t matrices, which should be replaced by the p functions in LMTO-ASA formalism. Since all the formulas below are valid in both LMTO-ASA and KKR cases, we prefer a more general notation. Three Pauli matrices

$$\boldsymbol{\sigma}_x = \begin{pmatrix} 0 & 1 \\ 1 & 0 \end{pmatrix}, \quad \boldsymbol{\sigma}_y = \begin{pmatrix} 0 & -i \\ +i & 0 \end{pmatrix}, \quad \boldsymbol{\sigma}_z = \begin{pmatrix} 1 & 0 \\ 0 & -1 \end{pmatrix}, \tag{5.64}$$

form a vector in the normal three-dimensional space.

The unit vector $\boldsymbol{e}$ defines the direction of a deviated spin. It equals ($\sin\theta$, 0, $\cos\theta$), and we are interested in its deviations from (0, 0, 1): $\delta\boldsymbol{e} = (\sin\theta, 0, \cos\theta - 1)$. Then,

$$\delta\boldsymbol{\tau} = \frac{1}{2}(\boldsymbol{\tau}_\uparrow - \boldsymbol{\tau}_\downarrow)(\delta\boldsymbol{e} \cdot \boldsymbol{\sigma}) = \frac{1}{2}(\boldsymbol{\tau}_\uparrow - \boldsymbol{\tau}_\downarrow) \begin{pmatrix} \cos\theta - 1 & \sin\theta \\ \sin\theta & 1 - \cos\theta \end{pmatrix}. \tag{5.65}$$

When this is substituted into (5.59), the trace over spin variables must be taken explicitly. Using the identity $\mathrm{Tr}\{\ln \boldsymbol{A}\} = \ln \det \boldsymbol{A}$, one obtains [5.43, 44]

$$J_0 = -\frac{1}{4\pi} \int^{E_F} dE \, \mathrm{Im}\{\mathrm{Tr}\{(\boldsymbol{\tau}_\uparrow - \boldsymbol{\tau}_\downarrow)(\boldsymbol{T}_\uparrow - \boldsymbol{T}_\downarrow) + (\boldsymbol{\tau}_\uparrow - \boldsymbol{\tau}_\downarrow)\boldsymbol{T}_\uparrow(\boldsymbol{\tau}_\uparrow - \boldsymbol{\tau}_\downarrow)\boldsymbol{T}_\uparrow\}\} \tag{5.66}$$

as a factor for θ^2 in the expansion of δE over θ (5.59). The trace in (5.66) is over orbital variables.

The possibility of calculating J_0 from first principles is crucial for the discussion of the stability of various magnetic phases. The self-consistent result of the usual band-structure calculation corresponds to some local or global minimum of the total energy. One has to keep in mind, however, that the magnetic moments in these calculations are usually allowed to change their magnitudes but not their direction. For most cases, this means that only collinear ferromagnetic or antiferromagnetic ordering is possible. Some recent calculations (discussed in Chap. 4) give an example of treating the non-collinear magnetic structures. However, the pattern of spin directions at various sites has still to be fixed before the calculation. In this case, it may be possible that the appearance of additional degrees of freedom related to free rotations of individual spins would lower the total energy as compared with what was erroneously believed to be the ground-state configuration. The effective exchange parameter J_0 shows whether this is the case. As is seen from (5.61). J_0 is positive for really stable configurations and negative for "false" solutions, i.e. for those magnetic configurations whose total energy can be lowered by deviating the spins from collinearity. The analysis of J_0 helps one to choose the real ground-state configuration when several of those are provided by the band-structure calculation, depending on the initial conditions.

Another important application of J_0 is related to the possibility of calculating magnetic phase transition temperatures and hence, in principle, of the magnetic phase diagrams. The weakest point here is the lack of well-defined prescriptions for choosing the quantum factor η in (5.62). It is obvious only that in metallic systems it should be chosen closer to unity than to the local-moments limit $(S+1)/S$ prescribed by the Heisenberg model. Uncertainty in this value may give an error of about 50% in predicting the Curie/Neel temperatures. This is, however, not as bad for studying the trends in the magnetic transition temperature for alloys, where the concentration is varied. In the coherent-potential approximation (CPA), the weighted exchange parameter for a substitutional binary alloy $A_{1-c}B_c$ is defined as

$$J_0(c) = (1-c)J_{\mathrm{A}}(c) + cJ_{\mathrm{B}}(c), \tag{5.67}$$

where $J_{\mathrm{A}}(c)$ and $J_{\mathrm{B}}(c)$ are effective exchange parameters for atoms A and B considered as impurities in the effective medium of a given concentration. The evaluation of the scattering path operator for a binary alloy is discussed in many papers devoted to the KKR-CPA formalism [5.25, 45–47]; what they say is mostly valid in the LMTO-ASA approach as well.

The slope of the $T_{\mathrm{C}}(c)$ curve at the ends of the magnetic phase diagram can be calculated without any reference to the CPA. It follows from (5.67) that

$$\left.\frac{dJ_0(c)}{dc}\right|_{c=0} = J_{\mathrm{B}} - J_{\mathrm{A}}, \tag{5.68}$$

where J_{B} is the effective exchange parameter for a single B atom embedded in the crystal of A, and J_{A} is the exchange parameter determined for the perfect A crystal.

Apart from the total exchange parameter J_0, the pair interactions J_{ij} constituting J_0 according to the sum rule

$$J_0 = \sum_{j \neq 0} J_{0j} \tag{5.69}$$

can also be calculated from the first principles. The values of these parameters are extracted from polarized neutron-diffraction measurements; they are positive or negative, according to whether the tendency for parallel or antiparallel ordering of magnetic moments within a given pair of atoms dominates.

To formulate the excitation $\delta\tau$ corresponding to this interaction, one needs to specify two atoms within a crystal and consider two magnetic moments deviating by, for simplicity, the same angle at two sites. In analogy to the method for J_0, one gets [5.44]

$$J_{ij} = \frac{1}{4\pi} \int^{E_F} dE \, \mathrm{Im}\{\mathrm{Tr}_L\{(\boldsymbol{\tau}_{i\uparrow} - \boldsymbol{\tau}_{i\downarrow}) \boldsymbol{T}^{ij}_{\uparrow}(\boldsymbol{\tau}_{j\uparrow} - \boldsymbol{\tau}_{j\downarrow}) \boldsymbol{T}^{ij}_{\downarrow}\}\}. \tag{5.70}$$

Deviations in the directions of one or two moments were considered above as examples of magnetic excitations, merely to specify the interactions whose parameters we wanted to calculate. In reality, the true basic excitations from the ground state of a ferromagnet are spin waves. The corresponding excitation energy (up to second order in $\boldsymbol{q}$) is

$$\delta E_{xc} = \sum_{\alpha\beta} D_{\alpha\beta} \delta \boldsymbol{q}_\alpha \delta \boldsymbol{q}_\beta, \tag{5.71}$$

where $D_{\alpha\beta}$ is the spin-wave stiffness tensor. To define it from first principles we should consider the non-local basic excitation, i.e. the deviation of spins in the whole lattice by the angles $\theta_i = \boldsymbol{q}\boldsymbol{R}_i$, where $\boldsymbol{R}_i$ is the lattice vector, and $\boldsymbol{q}$ is the spin-wave vector. It is convenient, then, to use the reciprocal-space representation in (5.59). The following expression for the spin-wave stiffness tensor has been obtained [5.47]:

$$D_{\alpha\beta} = \frac{1}{4\pi} \sum_{\boldsymbol{k}} \int^{E_F} dE \, \mathrm{Im}\left\{ \mathrm{Tr}_L \left\{ (\boldsymbol{\tau}_\uparrow - \boldsymbol{\tau}_\downarrow)^2 \frac{\partial \boldsymbol{T}(\boldsymbol{k})_\uparrow}{\partial \boldsymbol{k}_\alpha} \frac{\partial \boldsymbol{T}(\boldsymbol{k})_\downarrow}{\partial \boldsymbol{k}_\beta} \right\} \right\}, \tag{5.72}$$

where $\boldsymbol{T}(\boldsymbol{k})$ is the Fourier transform of the scattering path operator (5.31).

Back Fourier transformation to direct space provides the following expression:

$$D_{\alpha\beta} = \sum_{j \neq i} J_{ij} (R_i - R_j)_\alpha (R_i - R_j)_\beta. \tag{5.73}$$

For some applications, this formula can be more useful than (5.72), because the exchange parameters decrease rapidly as the corresponding interatomic distance grows. In [5.47], the calculation of the spin stiffness constant for metallic iron performed in the cluster approximation gave quite reasonable agreement with experimental values.

Another way to apply LSDA formalism to the problem of calculating the exchange parameters has been proposed [5.48–50]. The electronic structure of 3*d* metals at high temperatures was imitated using the so-called "alloy analogy" [5.46]. It means that the high-temperature paramagnetic state was treated within the same scheme as was developed for random binary alloys, i.e. the coherent-potential approximation. The main feature of the "alloy analogy" is that atoms with "up" and "down" directions of local magnetic moments are considered as constitutents of a binary alloy. Then, the coherent-potential approximation for the system in question can be formulated as

$$\tau_c^i = \tfrac{1}{2}(1+m_i)\tau_\uparrow^i + \tfrac{1}{2}(1-m_i)\tau_\downarrow^i + (\tau_\uparrow^i - \tau_c^i)T_c^{ii}(\tau_\downarrow^i - \tau_c^i). \tag{5.74}$$

where τ_c^i is the effective reverse scattering matrix for the site i in the paramagnetic alloy and T_c^{ij} is the effective scattering path operator. Both these properties correspond to the concentration c of atoms with spin-up moment; m_i is the average magnetic moment per site.

The effective exchange interaction parameter can be expressed as follows:

$$\begin{aligned} J_{ij} &= \frac{\partial^2\langle\Omega\rangle}{\partial m_i \partial m_j} \\ &= \frac{1}{\pi}\int^{E_F} dE\, f(E)\,\mathrm{Im}\left\{\mathrm{Tr}_L\left\{\sum_{k\neq i} D_\uparrow^i(\tau_\uparrow^i - \tau_\downarrow^i)D_\downarrow^i T_c^{ik}\frac{\partial\tau_c^k}{\partial m_j}T_c^{kj}\right\}\right\}, \end{aligned} \tag{5.75}$$

where Ω is the effective classical spin Hamiltonian determined within the LSDA approach. In (5.75), $f(E)$ is the Fermi-Dirac function, and $D_\sigma^i = [I + (\tau_\sigma^i - \tau_c^i) \times T_c^{ii}]^{-1}$. In the derivative of τ over m,

$$\frac{\partial\tau_c^i}{\partial m_j} = \tfrac{1}{2}D_\uparrow^i(\tau_\uparrow^i - \tau_\downarrow^i)D_\downarrow^i\delta_{ij} - \sum_{k\neq i} D_\uparrow^i(\tau_\uparrow^i - \tau_c^i)T_c^{ik}\frac{\partial\tau_c^k}{\partial m_j}T_c^{kj}(\tau_\downarrow^j - \tau_c^j)D_\downarrow^j, \tag{5.76}$$

the second term represents the effects of short-range spin ordering, which should be neglected in the coherent-potential approximation. This results in the following expression for the exchange parameter:

$$J_{ij} = \frac{1}{2\pi}\int^{E_F} dE\, f(E)\,\mathrm{Im}\{\mathrm{Tr}_L\{(\tau_\uparrow^i - \tau_\downarrow^i)\tilde{T}_\uparrow^{ij}(\tau_\uparrow^j - \tau_\downarrow^j)\tilde{T}_\downarrow^{ji}\}\}, \tag{5.77}$$

where $\tilde{T}^{ij} = D_\sigma^i T_c^{ij} D_\sigma^j$ defines the non-diagonal scattering matrix in the paramagnetic state. As is seen, (5.77) is actually similar to (5.70) if one keeps in mind that the non-diagonal elements of the scattering matrix for a ferromagnet are replaced with the elements of the effective T matrix of paramagnetic metal. The difference in the numerical factor is due to different definitions used for the basis for J_{ij}.

The Curie temperature has been calculated in [5.49] using the following equation:

$$T_C = \frac{1}{3k}\sum_j J_{ij}. \tag{5.78}$$

Note that for the case of a paramagnetic metal, it is not possible to perform the summation over the sites in (5.78) analytically; hence no simple formula for J_0 analogous to (5.66) can be derived.

A number of parameters important for practical applications are of essentially relativistic character, and the evaluation of the relativistic Green function is vital to calculate them from first principles. This does not refer only to compounds of heavy elements for which the importance of considering the relativistic effects is generally appreciated. Even for 3*d* transition elements, all properties related to the spin-orbit interaction and magnetovolume phenomena can be treated within the relativistic formalism only, since otherwise the spin direction is not coupled with any axes in the crystal lattice, and only the deviations of spins with respect to other ones can be discussed, as has been done above.

Thus, it is in principle clear how the one-electron problem can be solved in the relativistic case (including the evaluation of the Green function), but actual calculations are rather complicated and therefore not numerous. The main problems are the doubling of secular determinant size and the lowering of the symmetry if spin polarization is included on an equal footing with the relativistic effects. As a result, the one-site matrices which are diagonal in the ASA are of a general form in the spin-polarized relativistic calculation. With these complications, the main lines of solving the problem as discussed in Sect. 5.2 are still valid.

The practical aspects of performing a relativistic band-structure calculation and attempts to implement the spin-polarized relativistic calculation scheme have been discussed [5.51–54]. Calculations of the relativistic (non-magnetic) Green function have been performed [5.55–57] as a part of the KKR-CPA calculation. Calculations of the Green function for magnetic systems in the relativistic case are not known to us at the moment.

Once a calculation of the relativistic spin-polarized Green function has been performed, it will provide a basis for calculating further important parameters. An appropriate excitation to be considered can again be related to the deviation of local moments. The variation of the total energy in the relativistic case includes the spin-lattice coupling contribution, in addition to the exchange contribution of (5.59). Actually, these two contributions are heavily mixed together, and the simplest way of obtaining the net spin-lattice coupling is to fix the angles between the spins, i.e. to consider the uniform precession of spins in the lattice as an excitation. In contrast to the non-relativistic case, there is a non-zero contribution to the total energy variation already in first order over the deviation θ:

$$\delta E = \sum_{\alpha} V_\alpha \theta_\alpha + \sum_{\alpha\beta} K_{\alpha\beta}\theta_\alpha\theta_\beta . \tag{5.79}$$

where θ_α are the components of three-dimensional rotation vector $\vec{\theta}$ which now has an arbitrary direction and is not aligned along (100), as was supposed in (5.65). The expressions for V_α and $K_{\alpha\beta}$ follow from applying the same routine as before, i.e. using (5.59):

$$V_\alpha = \frac{\partial E}{\partial \theta_\alpha} = \frac{2}{\pi} \int^{E_F} dE \, \mathrm{Im} \{ \mathrm{Tr}_L \{ [\vec{t}^{-1} \times \vec{T}]_\alpha \} \}. \tag{5.80}$$

The vector notation for $\boldsymbol{t}$ and $\boldsymbol{T}$ means that they are expressed in terms of the Pauli matrices as follows:

$$\boldsymbol{t} = \boldsymbol{t}_0 + \vec{t}\vec{\sigma}. \tag{5.81}$$

The coefficients V_α are determined by the initial orientation of spins with respect to the crystal lattice; the condition $\vec{V} = 0$ defines possible easy magnetization axes. This condition is satisfied at least for high-symmetry directions in crystal. However, it is not a sufficient condition for the direction to be an easy magnetization axis, and an analysis of the second-order coefficient is crucial. If the system in question is an impurity embedded in crystal, then the condition $\vec{V} = 0$ determines possible directions of the impurity magnetic moment. In this case, $\boldsymbol{t}$ stands for the impurity-site scattering matrix (no longer diagonal in the relativistic formalism even for cubic crystals), and $\boldsymbol{T}$ is the scattering path operator for the impurity system.

The coefficients

$$\begin{aligned} K_{\alpha\beta} = \frac{1}{\pi} \int^{E_F} dE \, \mathrm{Im} \{ \mathrm{Tr}_L \{ & [(\vec{p} \cdot \vec{T})(\vec{p} \cdot \vec{T}) - \vec{p}\, T_0 \cdot \vec{p}\, T_0) - (\vec{p}\vec{T})] \delta_{\alpha\beta} + p_\alpha T_\beta \\ & - [\vec{p} \times \vec{T}]_\alpha [\vec{p} \times \vec{T}]_\beta + (\vec{p}\, T_\alpha \cdot \vec{p}\, T_\beta) - (p_\alpha T_\beta + p_\beta T_\alpha)(\vec{p} \cdot \vec{T}) \\ & + p_\alpha T_0 p_\beta T_0 \} \}. \end{aligned} \tag{5.82}$$

are related to the one-ion anisotropy tensor.

As analogous formula can be obtained for uniform spin precession in a perfect undisturbed crystal. Here the direct space $\boldsymbol{T}$ matrices of (5.82) should be replaced by their Fourier transforms $\boldsymbol{T}(\boldsymbol{k})$, and the sum over $\boldsymbol{k}$ has to be taken, along with the energy integration. The resulting tensor $K_{\alpha\beta}$ then defines the gap in the spin-wave spectrum that is an important property describing the true excitation from the ground state of a magnetically ordered crystal.

Another example of essentially relativistic interactions is the *Dzyaloshinsky-Moriya* antisymmetric exchange coupling [5.58]. Proceeding from its definition

$$\delta E_{as} = \sum_{ij} \vec{D}_{ij} [\vec{\theta}_i \times \vec{\theta}_j], \tag{5.83}$$

the following expression for its characteristic parameter can be obtained:

$$\vec{D}_{ij} = \frac{2}{\pi} \int^{E_F} dE \, \mathrm{Re} \{ \mathrm{Tr}_L \{ \vec{t}_i T_0^{ij} (\vec{t}_j \vec{T}^{ji}) - \vec{t}_j T_0^{ji} (\vec{t}_i \vec{T}^{ij}) \} \}. \tag{5.84}$$

In this chapter we have shown how calculation formulas for important parameters of magnetic interactions can be derived within the unified general approach. In the next chapters, we discuss the results of applying these formulas to various types of systems.

6. Results of Band-Structure Calculations for Transition Metals and Their Compounds

Theoretical models and computational approaches considered in previous chapters have been applied to the study of the electronic structure of such large classes of magnetic crystals that it is infeasible to review all of them. Instead, in the next two chapters we discuss some representative examples, which illustrate both the possibilities of direct numerical calculations of the electronic structure of magnetic crystals and give an idea of the results that can be obtained for different groups of crystals with different degrees of localization of magnetic electrons. The most delocalized are the itinerant states of magnetic electrons in metals.

6.1 Electronic Structure of Magnetic 3*d* Metals

Band-structure calculations of the magnetic structure of transition metals are by now quite numerous. A variety of computational results available make it possible to discuss the relative accuracy and shortcomings of various calculation schemes and provide a basis for better understanding of electronic properties of magnetic alloys. More recent calculations for ferromagnetic Fe, Co, Ni, and antiferromagnetic Cr and Mn give estimations of the equilibrium lattice constant, average magnetic moments, and cohesion energies that are in reasonable agreement with the experimental data.

High-precision calculations for bcc Fe have been performed [6.1–3] using the full-potential linearized augmented plane waves method (FLAPW).

Stability of the results for structural and magnetic properties of Fe with respect to the choice of various exchange-correlation potentials and to the accuracy of $\boldsymbol{k}$-space integration was specially studied. Figure 6.1 shows the curves of the total energy as a function of the lattice constant for non-magnetic and ferromagnetic bcc Fe, for cases of 30 and 90 $\boldsymbol{k}$ points taken in the irreducible wedge of the Brillouin zone, according to the results of *Hathaway* et al. [6.2]. The total energy is considerably higher for a more crude $\boldsymbol{k}$-points mesh, probably because of insufficient accuracy in describing small electron and hole orbits at the Fermi surface. Further increase in the number of $\boldsymbol{k}$ points from 90 to 210 hardly affects the total energy.

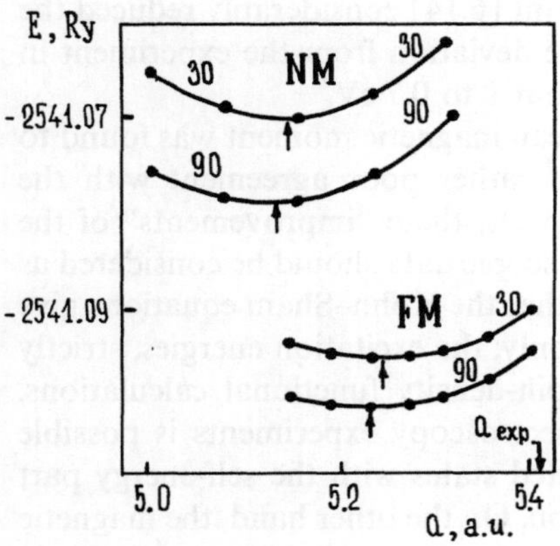

Fig. 6.1. Dependence of the total energy on the lattice constant value for bcc Fe, as calculated for non-magnetic (NM) and ferromagnetic (FM) phases with 30 and 90 k points in the irreducible wedge of the Brillouin zone [6.2]

The equilibrium lattice constant for ferromagnetic Fe obtained in the calculation is 3% smaller than the experimental value. This exceeds the estimated error of 0.1% in determining the equilibrium lattice constant; threfore, the error is probably introduced by the local approximation in the spin-density functional formalism used in the calculation.

This has been proved [6.4] using two different non-local density functionals in the LMTO calculations of total energies and related properties for several simple and transition metals. For ferromagnetic Fe, the estimates obtained for the lattice constant in high-precision LMTO calculations with the ***k***-space integration performed over 500 points in the irreducible wedge were 5.276 a.u. using the *Langreth–Mehl–Hu* non-local functional [6.5–7] and 5.445 a.u. using the *Perdew–Wang* non-local functional [6.8–10]. That is, in both cases the estimates were higher than in the local density approximation (5.203 a.u.) and rather close to the experimental value (5.406 a.u.). Another long-standing problem related to the calculations for ferromagnetic Fe – namely the fact that fcc Fe had a lower ground-state energy than the bcc phases – was proved by *Bagno* et al. [6.4] to be an artefact introduced by the local approximation for the density functional used in earlier calculations.

A comparison of data obtained in various papers for bulk properties of Fe shows that the magnitude of magnetic moment per atom is fairly well reproduced in almost all calculations, irrespective of the actual method of spin-density functional chosen. The values obtained vary from $2.18\mu_B$ [6.2, 11] to $2.15\mu_B$ [6.12], while the experimental value is $2.12\mu_B$. The remaining small systematical error is still to be explained in further studies.

A more striking discrepancy between theory and experimental results occurs for Ni, for which angle-resolved photoemission spectra are available. The *d* bandwidth is overestimated by about 30% in the calculation, and mean exchange splitting within the *d* band exceeds the experimental data by almost a factor of two [6.13]. A modified exchange-correlation potential with a variable degree of correlation effects taken into account [6.14] considerably reduced the *d* band spin splitting in the calculation. The deviation from the experiment in estimating the *d* bandwidth was reduced from 1 to 0.7 eV.

In this case, however, the value of the mean magnetic moment was found to be $0.53\mu_B$, which can be considered as in rather poor agreement with the experimental value of $0.606\mu_B$. In other aspects, these "improvements" of the calculation results as compared with spectroscopic data should be considered as successful only with some care. It is known that the Kohn–Sham equations give the correct description of the ground state only; the excitation energies, strictly speaking, cannot be extracted from the spin-density functional calculations. Correct comparison with photoelectron spectroscopy experiments is possible only on the basis of the calculation of excited states with the self-energy part taken into account in the local approximation. On the other hand, the magnetic moment is essentially a ground-state property, and its value is quite correctly reproduced in the usual LSDF band-structure calculations, e.g. those using the von Barth–Hedin exchange-correlation potential. The theoretical result for the Ni magnetic moment is $0.59\mu_B$ [6.12].

In ferromagnetic cobalt, which possesses a hexagonal structure, a considerable induced orbital magnetic moment occurs along with the pure spin magnetic moment. Large magnetocrystalline anisotropy related to this is typical for many Co-based compounds; for instance, the magnetocrystalline anisotropy of RCo_5 compounds is an order of magnitude higher than that of pure cobalt. As has been shown on the basis of band-structure calculation [6.15], both spin and orbital magnetic moments of Co are sensitive to the *c*/*a* ratio of a hexagonal lattice.

Figure 6.2 shows results of this analysis. The orbital magnetic moment which was estimated by perturbation theory [6.15] grows with the *c*/*a* ratio almost two times faster than the spin moment decreases. Therefore, the total magnetic moment increases.

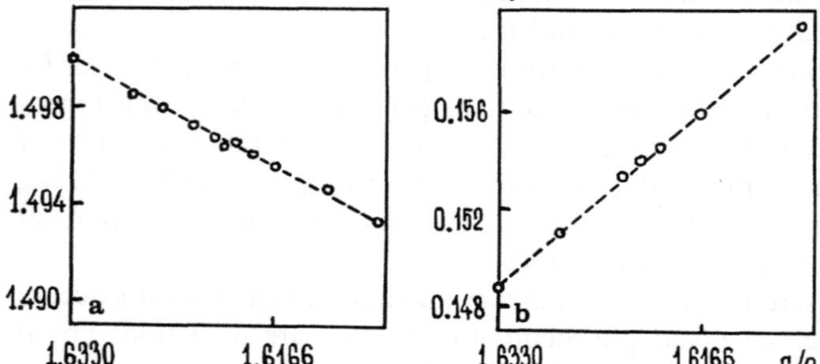

Fig. 6.2. Calculated values of (**a**) spin and (**b**) orbital magnetic moments of hcp Co as functions of the *c*/*a* ratio [6.15]

Calculated values of spin and orbital magnetic moments and g factors for 3*d* metals obtained in the LMTO calculation [6.16] are given in Table 6.1. Many observable anomalies in the magnetic and structural properties of transition-metal alloys are probably related to the complex dependence of the total energy on spin magnetization and interatomic distances in various lattice types. For example, the Invar behaviour of Ni–Fe alloys can be understood in terms of the interplay of low-spin and high-spin self-consistent ferromagnetic solutions occurring at various lattice spacings, as follows from the band calculation of fcc Fe [6.11].

The most detailed investigation of the band ferromagnetism of transition metals has been undertaken by *Moruzzi* et al. [6.17, 18]. They use the so-called "fixed spin moment" (FSM) scheme [6.19]. In this scheme, the mean magnetic moment of crystal M is kept fixed, along with the total number of valence electrons Z. These two parameters are determined as follows:

$$Z = N_\uparrow + N_\downarrow = \int^{E_F^\uparrow} dE\, n_\uparrow(E) + \int^{E_F^\downarrow} dE\, n_\downarrow(E), \tag{6.1}$$

$$M = N_\uparrow - N_\downarrow = \int^{E_F^\uparrow} dE\, n_\uparrow(E) - \int^{E_F^\downarrow} dE\, n_\downarrow(E). \tag{6.2}$$

An additional condition $M = \text{const}$ imposed on the system can be satisfied if the Fermi energies are allowed to be different for the two spin directions. This means that an external magnetic field $H = (E_F^\uparrow - E_F^\downarrow)/\mu_B$ is effectively introduced. The FSM method makes it possible to look for conditional minima of the total energy corresponding to various values of fixed magnetic moment M; thus one obtains more information than in the usual calculation when only the global energy minimum is searched for. Examples of extracting this additional useful information are given below.

Figure 6.3 shows the total energy E and the difference of the Fermi energies ΔE as functions of fixed spin moment M for bcc Fe [6.19]. The only minimum in the total energy curve corresponds to zero external field at the equilibrium magnetic moment value of $2.2\mu_B$. However, the situation can be more complicated in other systems, depending on the variation of the total energy with the interatomic distance.

Several types of $E(M)$ behaviour are schematically shown in Fig. 6.4. They are based on a general Landau expansion for the magnetic energy [6.20]:

$$E = E_0 + \alpha(V)M^2 + \beta(V)M^4, \tag{6.3}$$

Table 6.1. Experimental and theoretical values of spin and orbital magnetic moments and g factors of ferromagnetic metals

Metal	M_s^{calc}	M_L^{calc}	M^{exp}	g^{calc}	g^{exp}
bcc Fe	2.25	0.07	2.12	2.05	2.09
hcp Co	1.51	0.16	1.71	2.24	2.25
fcc Ni	0.62	0.04	0.60	2.1	2.18

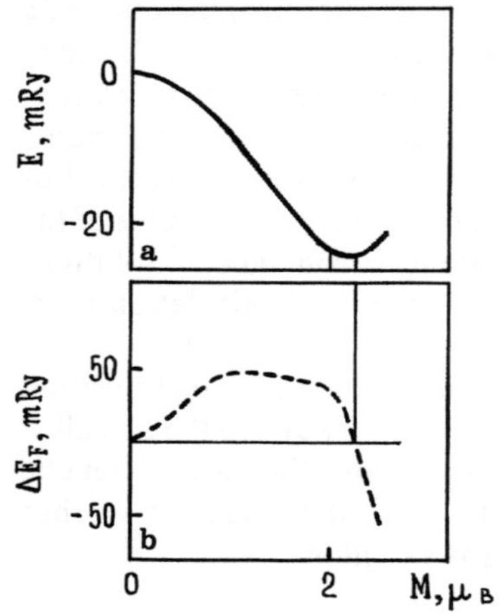

Fig. 6.3. (a) Magnetic energy E and (b) the difference of the Fermi energies $\Delta E_F = E_F^{\uparrow} - E_F^{\downarrow}$ calculated as functions of magnetic moment M for bcc Fe [6.19]

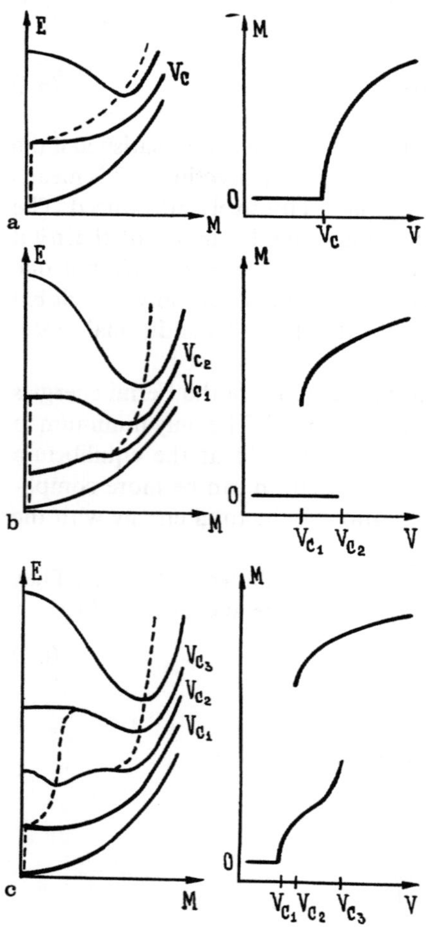

Fig. 6.4a–c. Three typical cases of evolution from a non-magnetic to a magnetic state. Dashed lines trace the shift of the total energy minimum

where E_0 is the total energy of the non-magnetic ground state of the crystal. The change from one type of behaviour to another occurs at critical values of volume V_C corresponding to spatial points in the α and β parameters. In all cases, the crystal is non-magnetic at sufficiently small interatomic distances, so that $dE(M)/dM = 0$, and $d^2E/dM^2 > 0$ at $M = 0$. On the other hand, for an expanded crystal $d^2E/dM^2 < 0$, and the total energy minimum occurs at non-zero magnetic moment.

There is an intermediate critical value of the lattice constant where $d^2E/dM^2 = 0$, so that for any larger values the equilibrium magnetic moment increases from zero to a finite number. The change of the equilibrim magnetic moment M, which corresponds to the minimum of the total energy (6.3), with the crystal volume is described by

$$M = \sqrt{-\frac{\alpha(V)}{2\beta(V)}}. \tag{6.4}$$

Near the critical volume V_c, α behaves as $\sim (V - V_c)$, and β can be considered to be a constant, in the first approximation. Then, $M \sim \sqrt{V - V_c}$, as shown in Fig. 6.4a.

More complex behaviour of the magnetic moment with lattice spacing can be found when two local energy minima coexist in the $E(M)$ curve within a certain range of volumes (Fig. 6.4b). At the first critical volume V_{c_1}, a local minimum at $M \neq 0$ starts forming, and at the second critical volume V_{c_2}, the non-magnetic solution ($M = 0$) ceases to be stable. In this case the system becomes magnetic, and hysteresis occurs in the dependence of the magnetic moment on volume.

An example of even more complex behaviour is shown in Fig. 6.4c, where two local minima are formed at non-zero magnetic-moment values. Here three critical volumes should be introduced, and there is a region in the $M(V)$ dependence where a low-spin or high-spin solution may occur, depending on the crystal's earlier stress treatment.

All types of magnetic behaviour described above have actually been found in band calculations of 3*d* metals that were performed for the fcc and bcc forms [6.18]. Figure 6.5 shows the actual $M(V)$ dependences obtained in these calculations. The volume is measured referring to the equilibrium volume (determined by the total energy minimum) for each metal.

The fictitious bcc Ni, which is non-magnetic at zero pressure, undergoes a transition into the ferromagnetic state at a volume only 1.5% larger than the equilibrium volume. The magnetic moment varies continuously in this transition; therefore, this transformation belongs to the first of the above-discussed types (Fig. 6.4a).

Fcc Co exhibits hysteresis, with a discontinuity in the magnetic moment, near a lattice volume reduced by 10% from equilibrium; this corresponds to the situation illustrated in Fig. 6.4b. Bcc V at very high (and therefore presumably not achievable) values of lattice expansion undergoes a complex metamagnetic

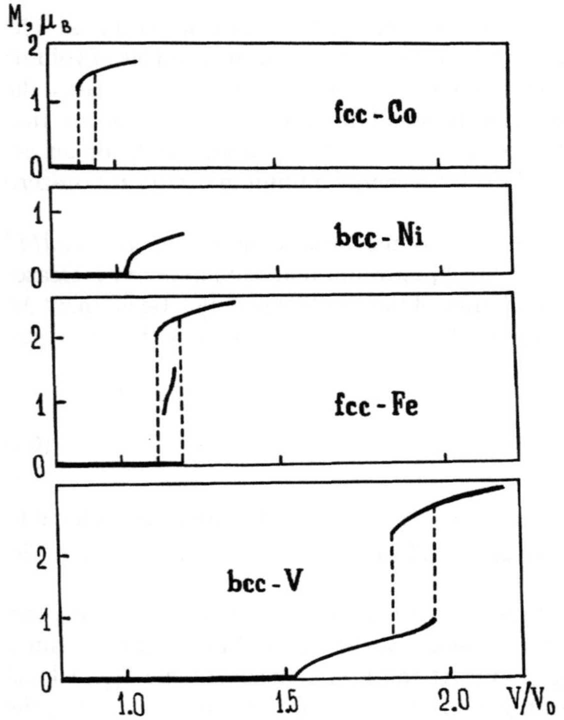

Fig. 6.5. Magnetic moment as a function of relative volume V/V_0 for several transition metals

transition defined by three critical volumes. In fcc Fe, any one of three possible solutions – namely, non-magnetic, low-spin on high-spin – may apply, depending on the actual volume once the lattice spacing has been increased by 20%. This complex behaviour characterized by four critical volumes is the result of the interplay of fine structure features of the density of states.

An important possibility provided by the FSM calculations is the opportunity to plot and analyse the total energy as a function of both lattice spacing and magnetic moment [6.17, 18]. Examples of contour maps describing these functions for bcc and fcc Fe [6.17] are shown in Fig. 6.6. Lattice spacings are expressed in terms of the Wigner–Seitz radii: $r_{WS} = a(8\pi/3)^{1/3}$ for the bcc lattice, and $r_{WS} = a(16\pi/3)^{1/3}$ for the fcc lattice.

In the usual band-structure calculations, only the total energy minimum and the magnetic moment corresponding to it are the results of calculation for each value of lattice spacing; therefore, only a vertical cut of the $E(M, V)$ surface along the deepest valley is studied. The global energy minimum corresponds to zero external pressure and zero magnetic field.

The energy surface for bcc Fe (Fig. 6.6a) has a global minimum at $M = 2.15\mu_B$ and $r_{WS} = 2.63$ a.u. – this agrees well with the experimental data. At all values of r_{WS} considered, the conditional energy minimum corresponds to non-zero

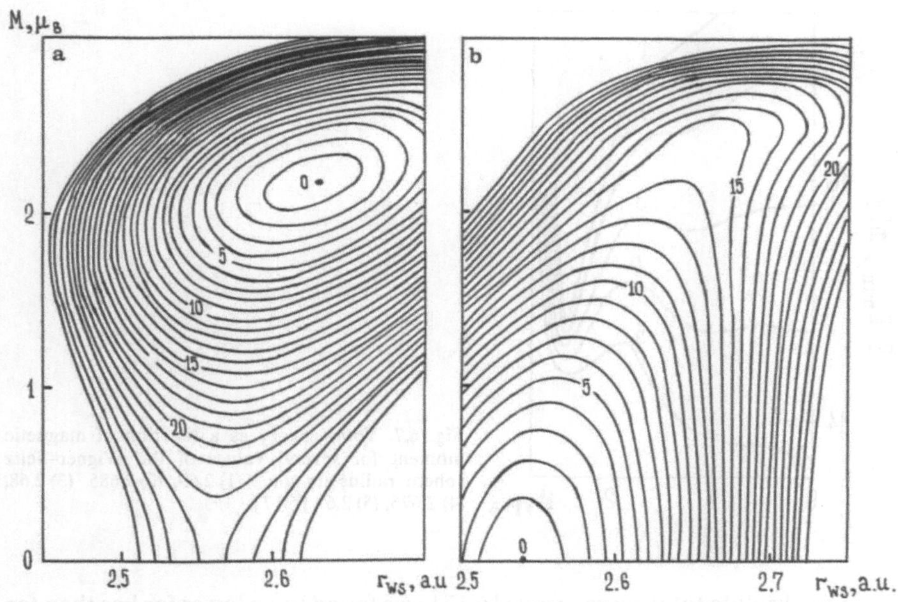

Fig. 6.6. Constant energy contours in the plane of M (magnetic moment) and r_{WS} (Wigner–Seitz sphere radius): (**a**) for bcc Fe, (**b**) for fcc Fe [6.17]

magnetic moments. Therefore, bcc Fe was found to be ferromagnetic at any reasonable pressure.

The $E(M, V)$ surface for fcc Fe (Fig. 6.6b) differs considerably from that for the bcc phase; it has a global minimum at $r_{WS} = 2.54$ a.u. and $M = 0$. Therefore, the ground state of fcc Fe is non-magnetic. However, a conditional total energy minimum at $M \simeq 1\mu_B$ appears at $r_{WS} = 2.68$ a.u.; then, already at $r_{WS} = 2.69$ a.u., the minimum jumps to $M = 2.4\mu_B$ and becomes quite pronounced. The magnetic moment corresponding to the energy minimum grows slightly further as the lattice spacing increases. *Bagayoko* and *Callaway* have studied the change of the magnetic moment in both fcc and bcc Fe for quite a broad range of lattice spacings [6.21].

The constant energy contours near the above-mentioned critical value of lattice spacing are practically orthogonal to the r_{WS} axis. This illustrates the fact that the magnetic moment changes abruptly within a very small interval of r_{WS} values.

This complex dependence of the total energy on r_{WS} and M is further explained by Fig. 6.7, which shows several cuts of the $E(M, V)$ surface along various r_{WS} = const lines. The Fe fcc phase can exist in a non-magnetic or high-spin ground state, depending on the (negative) pressure applied, or in a metastable low-spin state in the intermediate region, at the Wigner–Seitz sphere radius $r_{WS} = 2.68$ a.u.

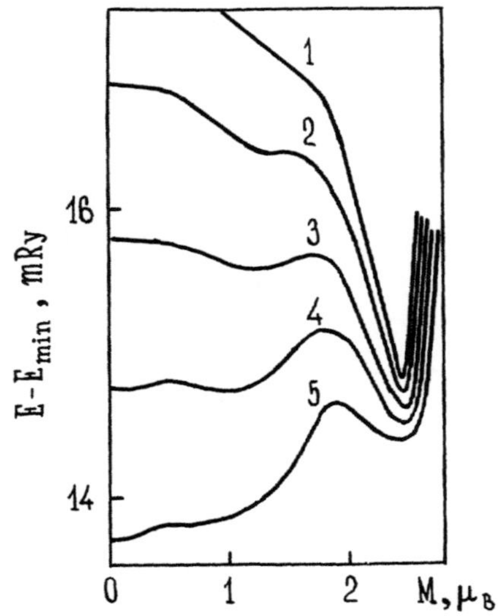

Fig. 6.7. Total energy as a function of magnetic moment for several values of the Wigner–Seitz sphere radius (in a.u.): (1) 2.69, (2) 2.685, (3) 2.68, (4) 2.675, (5) 2.67 [6.17]

The absolute total energy value [6.17] was found to be lower for bcc than for the fcc phase – consistent with the experimental evidence. This result is, however, by no means trivial, because in earlier papers [6.3, 11] the relation between the total energies of two phases was found to be opposite. Most probably, this drawback was due to the insufficient number of $\boldsymbol{k}$ points used in the tetrahedron integration over the Brillouin zone (40 $\boldsymbol{k}$ points in the irreducible wedge have been used [6.3]). When the $\boldsymbol{k}$-space integration was performed over 570 points in the irreducible wedge, it was possible to obtain the correct result [6.17].

When discussing the magnetic properties of fcc Fe, one has to take into account the possibility of forming the antiferromagnetic state. It has been supposed that the ground state of fcc Fe is antiferromagnetic, with magnetic structure of the CuAu type [6.11]. More precise calculations [6.3] performed by the FLAPW method showed that the ground state of fcc Fe is non-magnetic, as is consistent with the results of *Moruzzi* et al. [6.17]. However, as the lattice constant is slightly increased, the antiferromagnetic structure of fcc Fe (which was not considered in [6.17]) becomes more energetically favourable than both possible ferromagnetic states. At a further increase of interatomic distances, the high spin ferromagnetic structure becomes the ground-state one.

At first sight, the above-discussed predictions of the relative stability of various magnetic structures of fcc Fe are hardly of any practical interest, because the bcc-to-fcc transition temperature in Fe is higher than the Curie temperature. However, small crystals of fcc Fe can be stabilized as precipitates in the fcc crystal lattice of Cu or Au [6.22–25]. Depending on the host metal, the lattice spacing can be larger than in pure fcc Fe.

So far, only ordered magnetic structures of transition metals have been considered. Modern band-structure calculation methods make it possible to study the paramagnetic state (i.e. that with disordered local moments) of transition metals. This can be done on the basis of the so-called *alloy analogy*. The alloy analogy is the generalization of the calculation schemes created for binary substitutional alloys for the study of crystals consisting of chemically equivalent atoms, but with different orientations of their magnetic moments [6.26]. The method of calculating the effective exchange integral J_0 in the "disordered local moments" scheme was described in Sect. 5.3 Below, we give some calculation results for J_0 and other bulk properties obtained by the KKR method in the coherent-potential approximation for fcc Fe [6.27, 28].

Figure 6.8 shows calculated values of local magnetic moments in the "disordered moment" state of fcc Fe for various lattice spacings [6.28]. These results seem to be the most physically reasonable for describing magnetic properties of the high-temperature γ phase of pure Fe. To take into account the effect of temperature on the electron spectrum, the thermal broadening of the Fermi–Dirac distribution function corresponding to $T = 1400$ K has been introduced [6.28]. For comparison, Fig. 6.8 also shows the calculated values of magnetic moments in ordered ferro- and antiferromagnetic phases of fcc Fe. The $M(V)$ dependence at large lattice constant values in the disordered state is much closer to that for the antiferromagnetic configuration than for the ferromagnetic one. However, at $a = 6.6$ a.u. (which corresponds to $r_{WS} = 2.58$ a.u.) and smaller lattice spacings, the magnetic moment in the disordered state drops to zero.

More information on the interatomic magnetic interactions can be extracted from analysis of the static magnetic susceptibility $\chi(\boldsymbol{q}, T)$, which accounts for the effect of both Stoner-type one-electron transitions and multiparticle Heisenberg excitations. The $\chi(\boldsymbol{q})$ functions for $T = 1400$ K along the [111] direction as calculated by *Staunton* et al. [6.29] for fcc Fe are shown in Fig. 6.9 for several values of the lattice constant. The asymptotic value $\chi(\boldsymbol{q} = 0, T \rightarrow \infty) = M^2(V)/3kT$ – that is, the susceptibility of the system of non-interacting

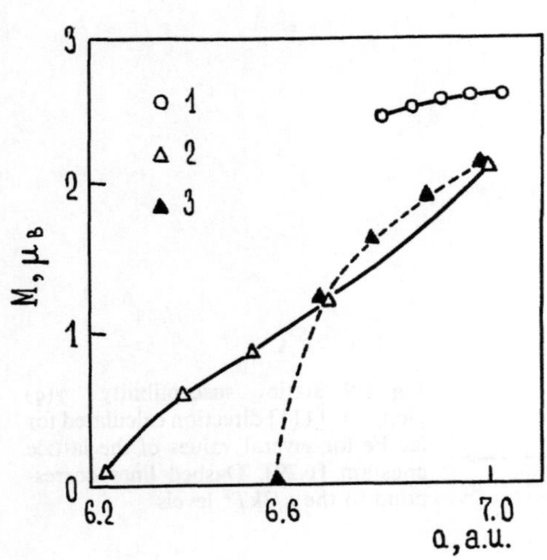

Fig. 6.8. Magnetic moment M as a function of lattice constant a for (1) ferromagnetic, (2) antiferromagnetic and (3) disordered magnetic states of fcc Fe [6.28]

electrons – is shown in Fig. 6.9 as a dashed line associated with each curve. The excess of $\chi(\boldsymbol{q})$ over this asymptotic value reveals the instability of the spin system towards the formation of the spin-wave state with given $\boldsymbol{q}$. A pronounced maximum of $\chi(\boldsymbol{q}, T)$ at $\boldsymbol{q} \simeq 0$ for $a = 7$ a.u. indicates that the magnetic structure of expanded fcc Fe is predominantly ferromagnetic. As the temperature decreases, the static susceptibility at $\boldsymbol{q} = 0$ becomes higher and diverges at the Curie temperature. At lower values of the lattice constant, the maximum of the static susceptibility shifts to non-zero $\boldsymbol{q}$ values. This means that a tendency to form an antiferromagnetic spin-wave-like state appears. Analysis of the static susceptibility along various directions in the Brillouin zone [6.28] has led to the conclusion that the absolute maximum of $\chi(\boldsymbol{q}, T)$ is achieved in the [111] direction. At the equilibrium lattice constant value of fcc Fe corresponding to room temperature ($a = 6.7$ a.u.; $r_{WS} = 2.62$ a.u.), the maximum occurs at the Brillouin-zone boundary $q = \pi/a$ [111]. This means that the ground-state antiferromagnetic structure is commensurate. This antiferromagnetic ordering is of the same type as in NiO, where Ni atoms also form a fcc structure. Note that the above-mentioned band-structure calculations of antiferromagnetic fcc Fe [6.3, 11] assumed a magnetic structure with $\boldsymbol{q} = \pi/a$ [100], i.e. that of the CuAu type.

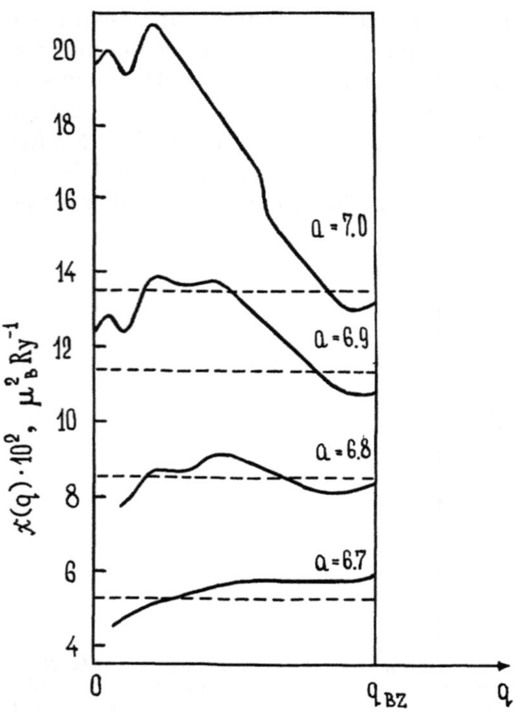

Fig. 6.9. Static susceptibility $\chi(\boldsymbol{q})$ along the [111] direction calculated for fcc Fe for several values of the lattice constant [6.29]. Dashed lines correspond to the $\mu/3kT^2$ levels

Two basically different methods have been proposed and implemented for calculating the Curie temperature in the ab initio LSDF-based approach. One of them starts from the fact that static paramagnetic susceptibility diverges at $T = T_C$, i.e. $\chi^{-1}(0, T_C) = 0$. This property for bcc Fe has been calculated by the KKR–CPA method [6.27], and the Curie temperature obtained (1280 K) is in reasonable agreement with the experimental value (1040 K).

Another method, which is simpler for practical applications, proceeds from the calculation of the effective exchange interaction parameter J_0 as defined by (6.66), with subsequent use of the mean-field approximation to estimate the Curie temperature by (6.62). The calculations performed by the LMTO–Green function method for bcc Fe revealed that J_0 is much more sensitive to the number of $\boldsymbol{k}$ points used in the calculation than the magnetic moment is. The J_0 value extrapolated to an infinite number of $\boldsymbol{k}$ points resulted in the estimate of $T_C = 1300$ K, consistent with the results of the above-mentioned susceptibility analysis.

Another advantage of the second approach is that it can be applied to study disordered substitutional alloys, where it is not possible to calculate explicitly the paramagnetic susceptibility.

Ab initio calculations of the electronic structure of antiferromagnetic metals (Cr, Mn) are not as numerous or profound as those for ferromagnetic metals. The main problem is the complexity of magnetic structure. In manganese, 58 atoms of the magnetic unit cell are distributed over four non-equivalent types of positions [6.30]; the magnetic ground state of pure chromium is described by a non-commensurate spin-density wave with the wave vector $\boldsymbol{Q} = 0.952 \cdot 2\pi/a$ [100] (see the review [6.31]). In the calculations so far performed, simplified structures have been considered, namely, the CsCl-type antiferromagnetic structure typical for the high-temperature δ phase Mn [6.32], and the same structure for antiferromagnetic Cr, with the deviation of the spin spiral from commensurability neglected [6.33–35].

In *Kübler*'s calculation [6.33], fairly good results as compared with the experimental data have been obtained for the equilibrium lattice constant and cohesion energy, and the calculated value of the magnetic moment per atom in the commensurate structure coincides with the experimentally known amplitude of the incommensurate spin wave ($0.59\mu_B$). It is important that the total energy of the antiferromagnetic state, according to the calculation, is ~ 0.06 eV lower than that of the paramagnetic state. The magnetic moment value turned out to be rather sensitive to interatomic distance. As *Kulikov* et al. noted [6.35], a decrease of the lattice constant by several percent from the experimental value results in an instant drop of the magnetic moment from $0.593\mu_B$ to $0.3\mu_B$.

Calculation of the electronic structure of paramagnetic Cr in the "disordered local moment" state [6.26] led to the collapse of the magnetic moment as a result of the self-consistent procedure. This means that the magnetism of chromium is essentially itinerant, and magnetic moments of individual atoms cannot survive in the absence of definite magnetic ordering.

6.2 Intermetallic Compounds and the Concept of Covalent Magnetism

The traditional view of the magnetism of metals and alloys supposes that a more or less clear distinction can be made between itinerant magnetic systems and systems with localized magnetic moments. In reality, one or another of these limiting cases is seldom encountered among intermetallic compounds.

Truly itinerant magnetic systems in the sense assumed in the Stoner model are not numerous; the Laves phase compound $ZrZn_2$ can be considered as an almost unique example. Note that none of the constituents of this intermetallic are magnetic materials themselves nor do they account for magnetism in other compounds. Nevertheless, $ZrZn_2$ is a ferromagnet because the Fermi level crosses a pronounced peak of the density of states, as was shown in the LMTO band-structure calculation by *Jarlborg* and *Freeman* [6.36]. They found that the product $In(E_F)$ which enters the Stoner criterion (3.11) exceeds unity by merely 0.01. All electron states in the valence band contribute to this value, creating the condition for the spin splitting of bands in energy by ~ 1 mRy, which results in a magnetic moment of $0.14\mu_B$ per unit cell. Taking into account such a low value of exchange splitting, one can consider $ZrZn_2$ as a true Stoner-type magnetic system, for which the spin-up and spin-down subbands are rigidly split as the internal magnetic field is introduced.

Rigid-band splitting as a way to lower the total energy during the transition to the ferromagnetic state actually occurs mostly in systems for which the Stoner critical value (3.11) is only slightly exceeded. For large values of $In(E_F)$, the simple splitting usually cannot ensure sufficient lowering of the DOS at the Fermi level and hence of the total energy. More radical changes of the band structure which establish ferromagnetism are usually accompanied by charge transfer and do influence the strength and spatial configuration of chemical bonds. It is a very energetically favourable situation for a system to possess either a majority-spin d subband completely filled, or a minority-spin subband completely empty. These cases arise quite often as a result of this transformation, and are usually referred to as strong ferromagnetism. The opposite case, with spin-up and spin-down d states partially filled, is called weak ferromagnetism. In considering the classification of real system with respect to this criterion, the concepts of magnetic valence and the generalized Slater–Pauling curve introduced by *Williams* et al. [6.37] are of much help.

The usual Slater–Pauling curve that gives the mean magnetic-moment value as a function of electron number per atom [6.38] is a broken line which ascends from the beginning of the $3d$ row and descends towards the end. Many experimental data for magnetic moments of $3d$ alloys and intermetallics lie on this curve, but there are several remarkable deviations which seem to be systematic and to obey some other dependence. *Williams* et al. [6.37] proposed using another property, namely, the magnetic valence

$$Z_m = 2N_{d\uparrow} - Z, \tag{6.5}$$

as an argument variable for the "generalized" Slater–Pauling construction. In (6.5) $N_{d\uparrow}$ is the occupation of the spin-up d states – i.e. 5 for strong ferromagnets belonging to the second half of the 3d row – and Z is the electron number per atom. For elemental metals, Z_m is an integer number and equals 2, 1, 0, -1, -2 for Fe, Co, Ni, Cu and Zn, correspondingly. For intermetallics and alloys, one should take the weighted average of Z_m as an argument value for the generalized Slater–Pauling plot.

For a given alloy, the total number of valence electrons is fixed, and the total magnetic moment equals

$$M = 2N_{\uparrow} - Z. \tag{6.6}$$

Comparison with (6.5) shows that the magnetic moment is merely magnetic valence (6.5) plus a doubled small contribution to the number of spin-up states from the s, p bands that equals ~ 0.3 within good accuracy for all late transition metals, as follows from the calculations performed by *Moruzzi* et al. [6.12]. Therefore, magnetic moments of transition-metal alloys are expected to fall onto the following line:

$$M = Z_m + 0.6. \tag{6.7}$$

This line is called the generalized Slater–Pauling (GSP) curve. After its introduction [6.37], it was discussed in more details in subsequent papers [6.39–41].

Several experimental data and calculation results for the magnetic-moment values are mapped on the GSP curve in Fig. 6.10 [reproduced from Ref. 6.41, Fig. 1]. Our present discussion refers to the ascending line on the left side of Fig. 6.10. No dramatic deviations of experimental data from the GSP occur, even for systems which exhibit a systematic deviating behaviour in the usual Slater–Pauling plot, for instance CoCr alloys. The values for this system are mapped onto the regular GSP curve because the very definition of the magnetic valence includes a drop from $Z_m = -6$ for Cr to $Z_m = 1$ for Co. Remaining

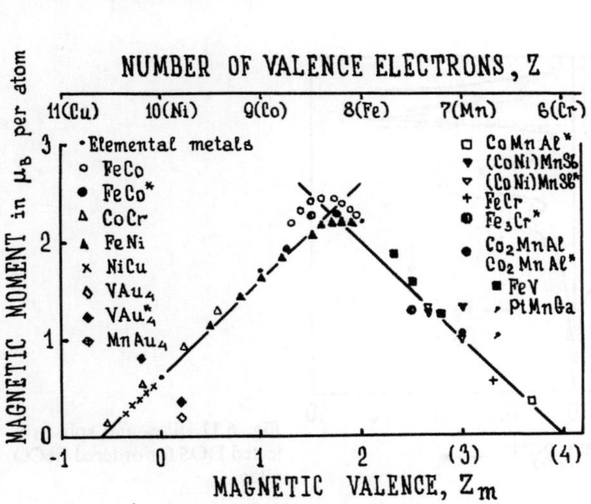

Fig. 6.10. Generalized Slater–Pauling curve [6.41]

deviations of experimental data from the curve which occur in Fig. 6.10 reveal important violations of the above-discussed simplified picture of strong ferromagnetism.

Experimental or calculated magnetic-moment values which are situated below the GSP curve correspond to systems which are weak ferromagnets. The intermetallic alloy VAu_4, the electronic structure of which has been calculated [6.41], is an example. Another and much better known case is pure Fe, which has a magnetic moment per atom equal to $2.2\mu_B$, and not $2.6\mu_B$, as it would be for a completely filled majority-spin *d* band. Also, Fe-based Ni and Co alloys are weak ferromagnets at small concentrations of the second component. The majority spin becomes fully populated with about 50% of Ni added to Fe, and the change of the magnetic moment on a subsequent increase of Ni concentration perfectly follows the GSP curve.

The FeCo system at 25 at % of Co is already a strong ferromagnet, and a subsequent increase of Co concentration enhances the magnetic moment above the values prescribed for strong ferromagnets by the GSP curve. The outstanding magnetic properties of the FeCo alloys were a subject of both experimental and theoretical considerations, among which the calculation by *Schwarz* et al. [6.42] was especially useful for clearing up the situation. The unusually high magnetic-moment value in the FeCo system arises as majority-spin states at both Fe and Co sites are filled at the expense of electrons removed from Fe minority-spin states.

The local DOS of Fe in Co in the ordered 50:50 FeCo alloy as calculated by *Schwarz* et al. [6.42] are shown in Fig. 6.11. The DOS of spin-up states is identical at both sites and has the shape typical for most bcc-structure metals. At the same time, the hybridization of spin-down *d* states of Fe and Co gives rise to bonding and antibonding subbands. The former are situated below the Fermi level, and the main contribution to them is from Co rather than Fe states, because the spin-down atomic *d* level of Co has lower energy than that of Fe. As

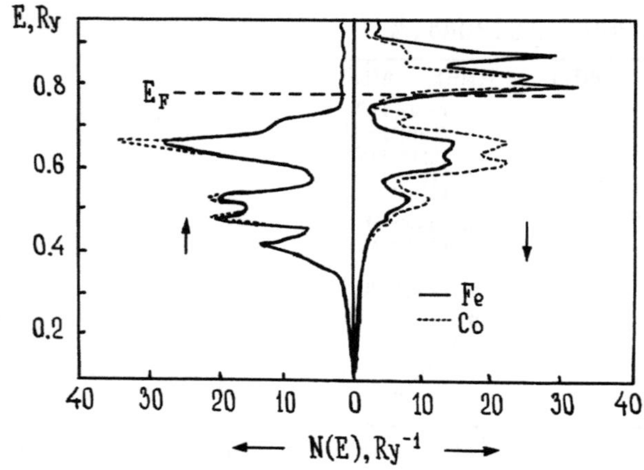

Fig. 6.11. Site- and spin-projected DOS for ordered FeCO alloy

a result, spin-down states at the Fe site are less populated than in pure bcc Fe, and the excess 0.3 electrons from these states rapidly fill the spin-up subbands at both Fe and Co sites as Co concentration grows, thus making the alloy a strong ferromagnet. The local magnetic moment at the Fe site reaches a maximal value of $2.72\mu_B$ at about 50% of Co, while at 25% of Co, maximal mean magnetic moment per atom ($2.30\mu_B$) is achieved. This is the highest value among all transition-metal-based systems, and is accompanied by slight charge transfer from Fe to Co (estimated at about $0.1e$ [6.42]). The absolute value of the charge transfer of course does not have much meaning, because it strongly depends on how it was defined, on the atomic sphere radii used in the calculation, etc. Nevertheless, the direction of the charge transfer as determined in the calculation should hardly be doubted. Note that the Fe-to-Co charge transfer direction is opposite to what should be expected from a rigid-band consideration, i.e. from the assumption that an additional electron provided by the Co atom is added to the common Co–Fe *d* band.

In contrat to FeCo alloys, the FeNi system exhibits the rigid-band-type filling of the *d* band as Ni concentration grows. This process is comparatively slow, and the alloy still remains a weak ferromagnet even at 25% of Ni [6.43].

Weak ferromagnetism does not allow any general treatment like that based on the GSP curve (6.7) in the case of strong ferromagnets. This is because the rate of filling spin-up and spin-down states in various rows of compounds depends strongly on the peculiarities of the chemical bonding in them. There are, however, many examples of systems in which the number of minority-spin electrons remains almost constant and roughly equal to 3 in a wide concentration range, irrespective of the population of majority-spin states. This is true for systems with a pronounced dip in the middle of the spin-down subband, as is, for example, typical for alloys with bcc structure. In this case the change of the magnetic moment with the electron-per-atom number is given by

$$M = Z - 6, \tag{6.8}$$

which defines the descending right-side part of the plot in Fig. 6.10. For alloys, Z denotes the average number of electrons per atom in the common valence band. The concept of magnetic valence is no longer useful in discussing alloys which are described by this part of the plot.

The magnetic moments of the FeCo and FeNi alloys at the Fe-rich side of the concentration scale – i.e. when the majority-spin subband is not yet completely filled – are in general better described by (6.8) than by (6.7). The effect of pinning the Fermi level in the dip of the minority-spin DOS in the FeCo system has already been discussed above. Even better agreement with (6.8) should be noted for bcc-structure binary alloys FeCr and FeV (Fig. 6.10).

Another broad class of systems for which the relation $N_\uparrow = 3$ is satisfied are Heusler alloys having the formula X_2MnY, where X = Co, Ni, Cu, Pd, etc., and Y = Al, Sn, In, Sb, etc. The electronic structure and magnetic properties of many such compounds have been calculated and discussed in detail [6.44]. A peculiar property of Heusler alloys is that density of minority-spin states has a deep minimum at the Fermi level, as follows from the calculation [6.44] for Co_2MnSn, Cu_2MnSn and Co_2MnAl, which are itinerant ferromagnets. As has been pointed out [6.41], the ferromagnetic compound Co_2MnSb even possesses a gap in minority-spin DOS at E_F, so that exactly 3 electrons are in spin-down states (Fig. 6.12). Therefore, magnetic moments of these Heusler alloys also obey (6.9) with good accuracy.

Two important questions arising in relation the the Heusler alloys are the degree of the magnetic-moment localization in them and the origin of definite magnetic structure (ferromagnetic or antiferromagnetic) in different compounds of this class. It has been established [6.44] that magnetic moments at Mn sites are localized, as was supposed already in earlier studies. But quite unexpectedly, these magnetic moments were found to be arising from completely itinerant electronic states. Majority-spin *d* states at Mn sites strongly hybridize with other constituent states and form a broad common valence band. So these states can be considered as more or less uniformly distributed over the crystal. At the same time, minority-spin states of Mn atoms are situated above the Fermi level and hence remain practically empty. The exclusion of the spin-down charge density from Mn sites results in the formation of localized magnetic moments. Figure 6.13 illustrates this effect.

Using the total energy calculations performed by *Kübler* et al. [6.44], it was possible to compare the relative stability of various magnetic phases (a ferromagnetic one and two different antiferromagnetic ones) for a number of Heusler alloys. The magnetic moments of Mn atoms were found to be practically insensitive to the type of magnetic ordering; this is related to their localized nature discussed above. Of all the compounds considered, only Pd_2MnIn has the lowest total energy in an antiferromagnetic configuration, namely with alternating planes of "positive" and "negative" magnetic moments being parallel to the (111) plane of the crystal. In this structure, antiferromagnetic coupling between adjacent Mn atoms is mediated by the In *p* states.

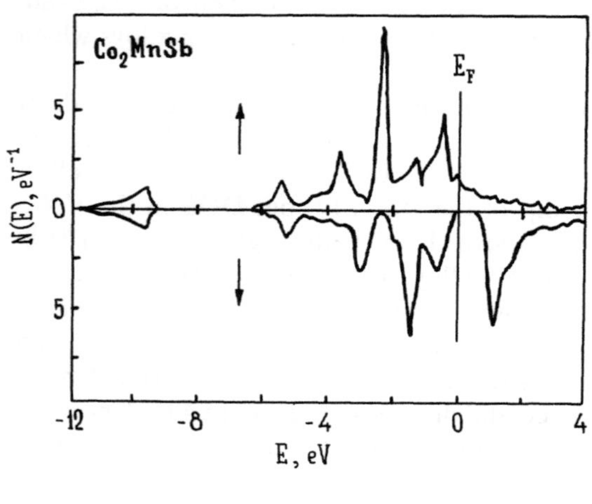

Fig. 6.12. Total spin-projected DOS of Co_2MnSb [6.41]

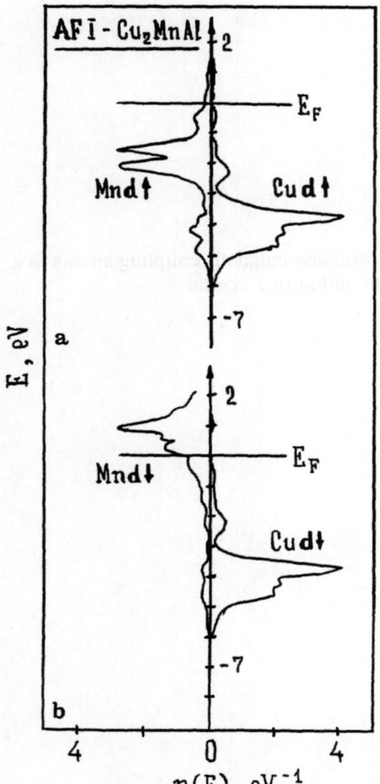

Fig. 6.13. Site- and spin-projected *d*-electron sublattice state densities for assumed antiferromagnetic Cu_2MnAl [6.44]

Chapter 2 discussed on a qualitative level, why antiferromagnetic ordering is preferential in some systems. Based on numerical results of ab initio calculations. *Moruzzi, Kübler, Williams* et al. stressed in a series of papers [6.43–46] the important effect of the spin-dependent part of covalent interaction on the competition between possible magnetic structures in alloys and compounds of transition metals. Figure 6.14 shows schematically the main effect of covalent mixture between spin-split *d* states of a transition-metal atom (site 1) and a non-magnetic neighbour which mediates the interaction (site 2). Let us suppose, for simlicity, that the spin-up state at site 1 prior to interaction is filled and the spin-down state is empty. As a result of the interaction, two low-energy bonding and two high-energy antibonding states appear. The total energy of the system can be considerably reduced if both bonding states are filled and two antibonding states are empty. The lowest spin-up bonding state cannot contain all electrons from site 1, and extra spin-up electrons should either populate high-lying antibonding states (without any energy gain), or tunnel to spin-up bonding states formed between the same intermediate atom and adjacent transition-metal centre. This "neighbouring" spin-up bonding state is not filled if the

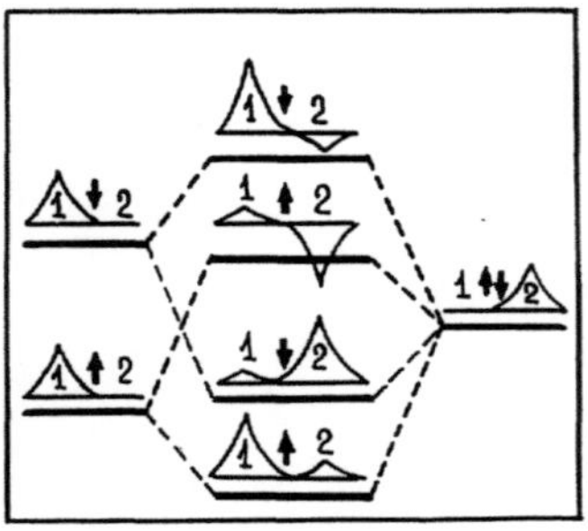

Fig. 6.14. Scheme of antiferromagnetic coupling arising as a result of covalent interaction in a crystal

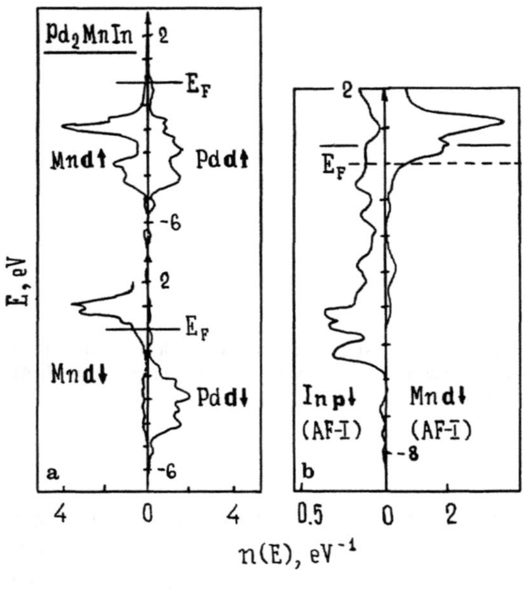

Fig. 6.15. Local densities of states in the Heusler alloy Pd_2MnIn [6.44]: (a) formation of localized magnetic moments at the Mn site, (b) hybridization of Inp and Mnd states, which results in antiferromagnetic coupling of Mn atoms

magnetic moment at the adjacent site is reversed, i.e. if the lowest spin-split state is the spin-down one. Conversely, excess spin-down electrons from the neighbouring centre tunnel to fill the next-lowest bonding state formed between site 1 and intermediate atom 2, and thus the problem of distributing electrons over low-energy bonding states is solved. This mechanism is especially efficient with a half-filled d shell. On the other hand, covalent interaction with non-magnetic states reduces the effective spin splitting at the transition-metal site, and the magnetic energy is reduced. A competition of these two tendencies in the total energy leads to an energy preference for ferromagnetic or antiferromagnetic ordering.

The covalent interaction between Mn d states and In p states in Pd_2MnIn manifests itself in a pronounced additional peak in both Mn $d_\downarrow$ and In $p_\downarrow$ DOS above the Fermi level (Fig. 6.15).

The comparison of total energies calculated for ferromagnetic and antiferromagnetic configurations made it possible to estimate the interatomic exchange integrals and Curie (Néel) temperatures of the Heusler alloys considered, based on simple formulas which follow from the Heisenberg model. For all alloys but one (Cu_2MnSn), an agreement with the experimental data in estimates of paramagnetic Curie temperature within 20% was achieved.

Various approaches described in this section proved useful in many situations for better understanding and discussion of the calculation results which follow from traditional self-consistent calculations performed within the usual LSDF approach. However, with semiconductors and insulators as subjects for calculation, it became clear that the LSDF itself is responsible for obvious errors in determining the bandwidths and the values of magnetic moments. A brief review of attempts to build an adequate band-structure description of such systems is given below.

6.3 Antiferromagnetic Monoxides

For a long time the transition-metal monoxides have been a challenge to solid-state physicists, because the microscopic origin of the metal-insulator transition in these systems, as well as the nature of their interatomic magnetic coupling, is still unclear [6.47–54]. The 3*d* monoxides of the beginning of the transition-metal row (TiO and VO) are non-magnetic metallic compounds, and their electronic structure has been successfully described in the ordinary band theory by *Mattheiss* [6.55, 56]. The monoxides of the 3*d* metals belonging to the end of the row are all antiferromagnetic insulators; it is commonly believed that these materials are Mott–Hubbard insulators, and the magnetic coupling in them is due to superexchange (Chap. 1). Most of the earlier theoretical studies of 3*d* metal monoxides proceeded from the strong correlation limit, i.e. $U/W \gg 1$, where U is the intra-atomic Coulomb integral, and W is the valence band width [6.49]. Recent analysis of the electronic structure and photoemission spectra of NiO-type compounds in terms of the Anderson lattice many-electron Hamiltonian have led to the following assumption: NiO is a charge transfer insulator where the charge transfer from O 2*p* to Ni 3*d* states takes place, and Δ, i.e. the characteristic energy related to it (Sect. 2.4), is smaller than U, the Coulomb interaction energy of the Ni 3*d* electrons. Nevertheless, later experiments on the angle-resolved photoemission spectra of NiO and CoO [6.57] made it clear that the lattice effects or interaction between 3*d* states is quite important even for localized 3*d* states of these compounds.

Extensive studies of the electronic band structure of the transition-metal monoxides in the local spin-density functional approach have been made by *Andersen* et al. [6.58] and *Oguchi* et al. [6.59–61]. They showed that the trends in the lattice parameter and, in particular, a sudden jump of the interatomic distance when going from VO to MnO which is magnetic in origin, can be reproduced in the band-structure calculations [6.58]. A more refined problem regarding the nature of the energy-band gap has also been discussed on the basis of band-structure calculations [6.61]. Figure 6.16 shows the calculation results for different types of magnetic ordering in MnO. The states below ~ 0.35 Ry are mostly of O 2*p* character and those lying above this energy are formed by the Mn 3*d* states in the following order: $t_{2g\uparrow}$, $e_{g\uparrow}$ and $t_{2g\downarrow}$, $e_{g\downarrow}$. In the ferromagnetic (F) state, there is no band gap at the Fermi energy. The main reason for the metallic behaviour of the F states is the relatively large width of the e_g partial subband, which is due to the O bonding, and small exchange splitting. Nevertheless, the exchange splitting is sufficient to produce a deep minimum in the total density of ferromagnetic states.

Two types of antiferromagnetic ordering, AF–I and AF–II, schematically shown in Fig. 6.17, have quite different total densities of states. AF–I has a DOS similar to that of the ferromagnetic state, while the AF–II state has a very narrow e_g band and a pronounced band gap about 0.1 Ry wide.

The paramagnetic state has been considered as a medium of randomly oriented moments treated within the Korringa–Kohn–Rostoker coherent-potential approximation (KKR–CPA) using the atomic-spheres approximation. The most interesting feature of this paramagnetic DOS is the presence of an

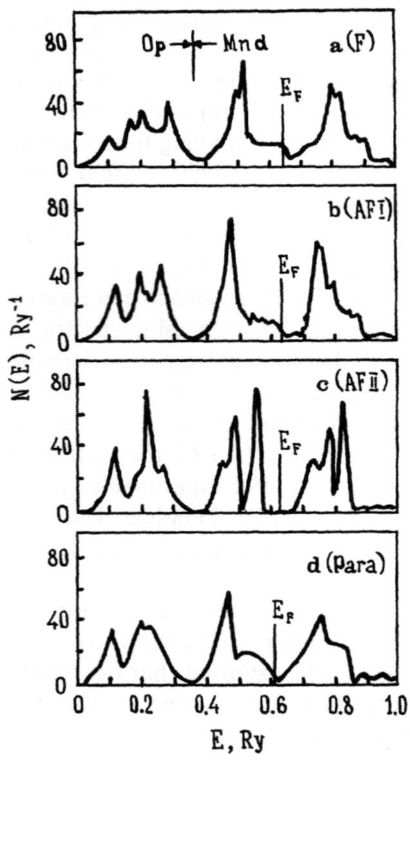

Fig. 6.16. Total densities of states of MnO in **(a)** the ferromagnetic state, **(b, c)** two different antiferromagnetic states and **(d)** the paramagnetic state in [6.61]

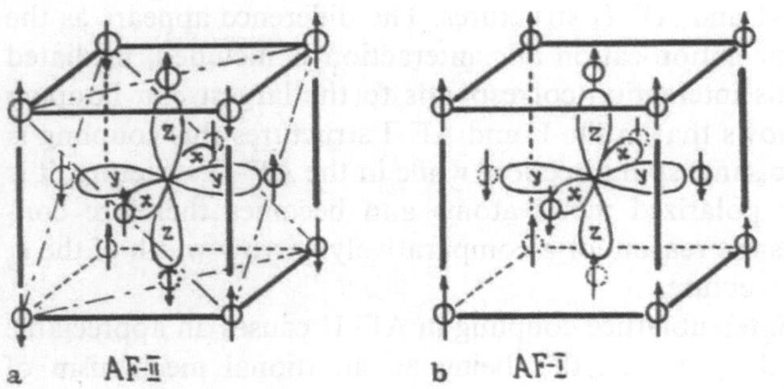

Fig. 6.17a, b. Two kinds of antiferromagnetic ordering considered in the band-structure calculations of transition-metal monoxides [6.61]. Circles with arrows denote 3d atoms; oxygen atoms are situated at the centres of the cubes and at the middle of the edges

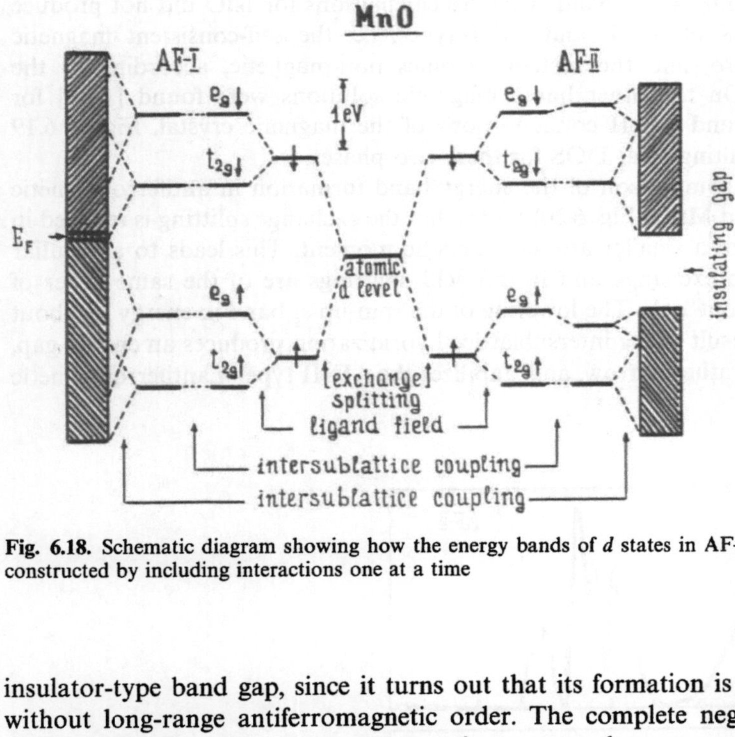

Fig. 6.18. Schematic diagram showing how the energy bands of d states in AF–II and AF–I are constructed by including interactions one at a time

insulator-type band gap, since it turns out that its formation is possible even without long-range antiferromagnetic order. The complete neglect of short-range magnetic order in the paramagnetic state was, however, a weak point in the cited calculation.

According to experimental evidence, the magnetic state of the 3d metal-insulator oxides is of the AF–II type. Figure 6.18 shows a scheme of the energy bands formed from the corresponding atomic levels, as different types of interactions are included, for both AF–I and AF–II structures. The largest splitting is of the exchange type (by ~ 0.28 Ry), then the ligand field-type splitting (induced by the interaction with 6 neighbouring oxygen atoms) of a 3d level into t_{2g} and e_g components (by ~ 0.08 Ry). Up to this point, there is no distinction between AF–I and AF–II structures. The difference appears as the second-nearest-neighgour cation-cation $dd\sigma$ interaction is included, mediated by the O p_z orbital. This interaction corresponds to the largest $dp\sigma$ hopping integrals. Figure 6.17 shows that in the F and AF–I structures this coupling is between the states of the same spin direction, while in the AF–II structure, it is between the oppositely polarized metal atoms and becomes therefore considerably weaker. This is the reason for a comparatively narrow width of the e_g subband in the AF–II structure.

An inclusion of the intersublattice coupling in AF–II causes an appreciable energy shift of occupied $e_{g\uparrow}$ states, this being an additional mechanism of forming the energy band and stabilizing this type of antiferromagnetic structure in MnO.

Similar LSDF-based band-structure calculations for NiO did not produce magnetic states of the F and AF–I types, i.e. the self-consistent magnetic moment is zero, and the system becomes non-magnetic, according to the calculations. On the other hand, magnetic solutions were found [6.61] for paramagnetic and AF–II configurations of the magnetic crystal. Figure 6.19 shows the resulting total DOS for these two phases.

Schematic comparison of the energy-band formation in antiferromagnetic oxides NiO and MnO (Fig. 6.20) shows that the exchange splitting is reduced in NiO because of a smaller atomic magnetic moment. This leads to a peculiar situation when exchange and ligand-field splittings are of the same order of magnitude, about 1 eV. The lowering of the spin-up e_g band in energy by about 0.04 Ry as a result of the intersublattice hybridization produces an energy gap, although it is rather narrow, and stabilizes the AF–II type of antiferromagnetic ordering.

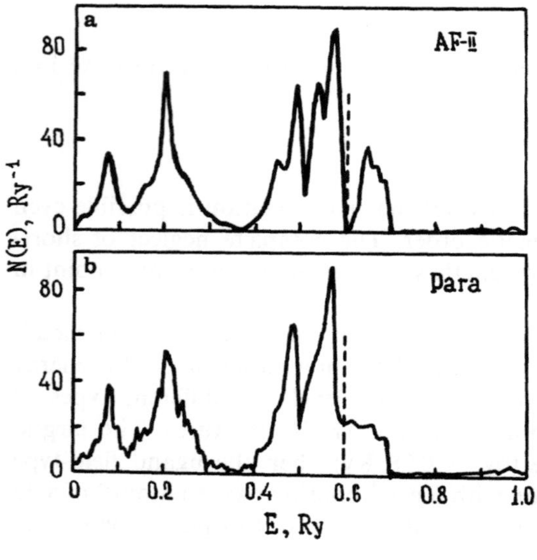

Fig. 6.19. Total DOS of NiO for (**a**) the AF–II state and (**b**) the paramagnetic state [6.59]. The dashed line indicates the Fermi level

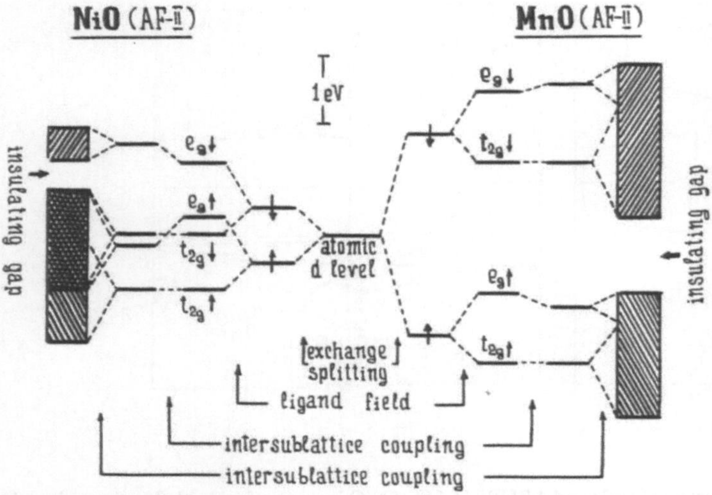

Fig. 6.20. Schematic diagram showing how the energy bands of *d* states are constructed in NiO and MnO with AF–II structure [6.61]. In MnO, the exchange splitting dominates over other energy scales, while in NiO, the exchange splitting, ligand-field splitting, and *d*-band width are all comparable in magnitude

In the KKR–CPA calculation for the paramagnetic phase of NiO, the bands are broadened and the band gap at the Fermi level disappears. It contradicts the experimental evidence that NiO, as well as MnO, is an insulator at temperatures above the Néel point. This means that correlation effects beyond those already included in the local spin-density approximation may be important for these systems.

The values of screened Coulomb interaction parameters U have been evaluated in a supercell calculation [6.62]. The resulting values of $U = 4$ eV led to the conclusion that an estimate of the energy of both the *d*-electron and *d*-hole excitation should be increased by ~ 2 eV, as compared with the result of the LSDF calculation, thus leading to an increase in the energy gap in NiO from 0.6 eV to 4.6 eV. Comparison of the resulting DOS in NiO with experimental bremsstrahlung isochromate spectra (BIS) [6.63] and the angle-resolved ultraviolet photoelectron spectrum (ARUPS) [6.64] in Fig. 6.21 shows their good agreement with the "model" band structure resulting from calculation [6.62]. Moreover, the recent angle-resolved photoemission study of NiO [6.57] agrees very well with the calculation results with respect to the O 2*p* band dispersion (however, the energy separation between the O 2*p* and Ni 3*d* bands should be reduced by 2.15 eV to fit the experimental data). Therefore, an adequate calculation of the electronic structure of magnetic oxides should include a mechanism for displacement of the metal 3*d* states due to effect of the Coulomb interaction, thus going beyond the LSDF.

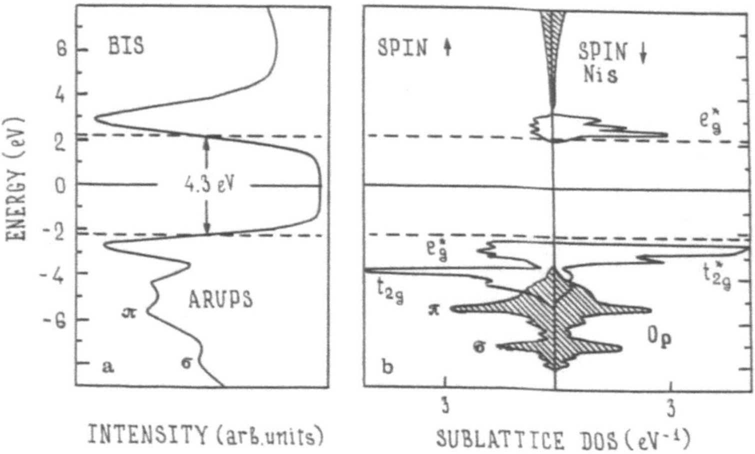

Fig. 6.21. (**a**) BIS spectrum [6.63] and ARUPS [6.64] of NiO in comparison with (**b**) spin-projected DOS of antiferromagnetic NiO [6.62]. Shaded portions of the DOS are as calculated; states labelled with an asterisk are shifted by the calculated Coulomb energy U

A method proposed by *Anisimov* et al. [6.65] to introduce a correction to the potential of unoccupied d states that effectively moves them upwards from the occupied ones improved the calculated values of the band gap compared with experimental data and also resulted in better estimates of local magnetic moments in the $3d$ oxides. These two properties (band gap and magnetic moment) are closely related because the unoccupied orbitals in MnO, FeO, CoO and NiO are only of the spin-down type. Their shift in energy away from the O $2p$ band lowers the occupation of spin-down states only, thus increasing the local magnetic moment at metal sites. In Fig. 6.22, the calculated (within the LSDF) values of local magnetic moments and of the charge transfer from Ni to O atoms in NiO are shown as functions of atomic-spheres radii, with the lattice constant kept fixed. The magnetic moment is seen to be rather sensitive to the choice of the sphere radius, and rather unexpectedly it becomes smaller as a larger Ni atomic sphere is taken.

We believe this is a drawback of the calculation scheme and can be understood as follows. In reality, the spin polarization is non-uniform not only on the interatomic scale but also inside the atomic spheres of both kinds of atoms. If traced along the Ni–O–Ni line, the spatial spin polarization would change its sign near the oxygen atom as compared with its behaviour near the nickel atom. With the atomic-spheres approximation used by *Terakura* et al. [6.61], the potential is spherically averaged within the atomic cells; therefore, the net polarization at oxygen sites is zero. The above-mentioned non-uniformities in the spin polarization can be taken into account only if they are included in the atomic spheres of surrounding nickel atoms. Therefore, the bigger the atomic sphere chosen around the Ni atom, the more compensating spin polarization

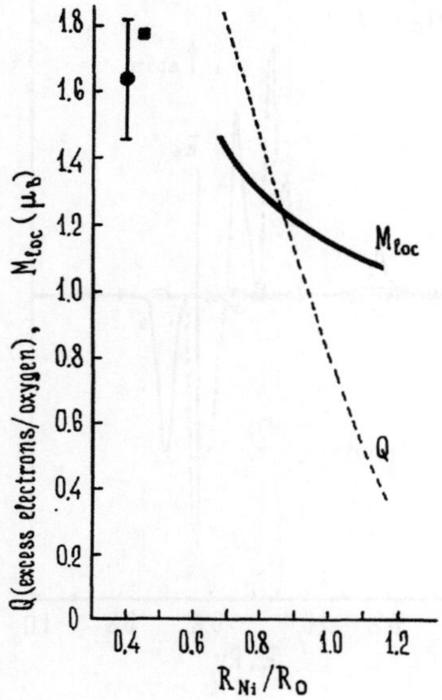

Fig. 6.22. Spin magnetic moment within the atomic sphere of Ni in NiO, M_{loc}, and the electron transfer from Ni to O, Q, as functions of the atomic-spheres radii R_{Ni}/R_O. The experimental data for the Ni magnetic moment are shown by a square and a circle with an error bar [6.61]

from neighbouring oxygen atoms enters it, thus lowering the resulting net spin moment at Ni sites. With the usual prescription for determining the atomic sphere radius, the calculated magnetic moment equals $1.09\mu_B$, while the experimental result for the Ni moment in NiO is $1.77\mu_B$ [6.61]. Effectively taking into account the Coulomb correlation effects [6.65] enhances the magnetic moment up to $1.6\mu_B$, which is in considerably better agreement with the experimental data.

The situation concerning the LSDF band-structure calculations for FeO and CoO seems to be even more dramatic. Even in the AF–II ground state, no band gap at the Fermi level was found in the calculations. Figure 6.23 shows the partial density of t_{2g} and e_g states in FeO and CoO according to *Terakura* et al. [6.61]. The Fermi level crosses the spin-down t_{2g} band, which has a fairly high density of states, thus predicting a metal-type electronic structure of FeO and CoO, in contrast to the experimental data, according to which both compounds are insulators. We believe that including the correlation effects beyond the local density functional theory would lead to a splitting of the $t_{2g\downarrow}$ state. An important related feature is the presence of a fairly large unquenched orbital moment in both materials. This leads to instability of the population balance in the $t_{2g\downarrow}$ bands induced by the intra-atomic exchange interactions. In the Anderson model, the condition for the existence of a non-zero orbital moment [compare (3.45)] is given by

$$N(E_F)(U-J) > 1, \tag{6.9}$$

where $N(E_F)$ is the density of states per spin and per orbital at the Fermi level,

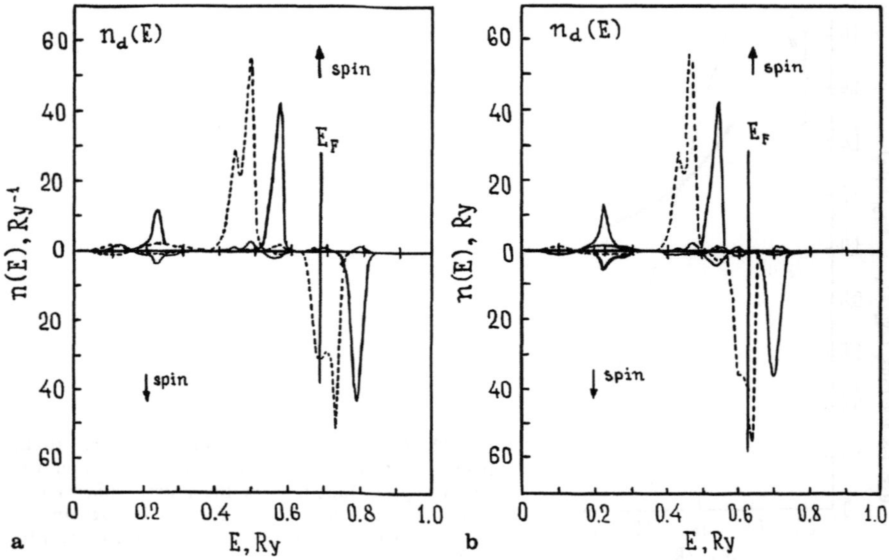

Fig. 6.23. Partial *d* DOS of transition-metal atoms in (**a**) FeO and (**b**) CoO as calculated without population imbalance in the t_{2g} band of the minority-spin state [6.61]. The solid (broken) lines denote $e_g(t_{2g})$ contributions

and U and J are effective Coulomb and exchange integrals for fivefold degenerate *d* orbitals. Using experimental estimates of $U \simeq 3$ eV and $J \simeq 1$ eV for the 3*d* transition metals [6.66] and the calculated values of $N(E_F)$, *Terakura* et al. obtained [6.61] $N(E_F)(U - J) \simeq 2.5$ for CoO and 1.8 for NiO.

Therefore, FeO and NiO are unstable with regard to the population imbalance in the $t_{2g\downarrow}$ band, which indicates the possibility of a gap opening up at the Fermi level. It is possible also, with the same values for U and J, to check the validity of the Stoner criterion for instability with respect to the formation of the local spin moment for the transition-metal oxides [similar to the criterion appearing in the Anderson model as (3.44)]:

$$N(E_F)(U + 4J) > 1. \tag{6.10}$$

This criterion is also satisfied for FeO and CoO, but the LSDF estimate of $(U + 4J)$ as $5I$, where I is the effective Stoner parameter, is approximately 40% smaller than the "experimental" value. It is possible to consider a simple degenerate Anderson impurity model, which allows a population imbalance only at a given Fe or Co site [6.61]. Figure 6.24 shows the results of this model analysis for FeO and CoO with the values of U and J experimentally determined and the density of states obtained from the band-structure calculation. The $t_{2g\downarrow}$ subband splits into two almost separate bands, resulting in a very small density of states at the Fermi level, i.e. a predecessor of the band gap. Of course, it is not possible to imitate a real band gap in the density of states at one site if there is no

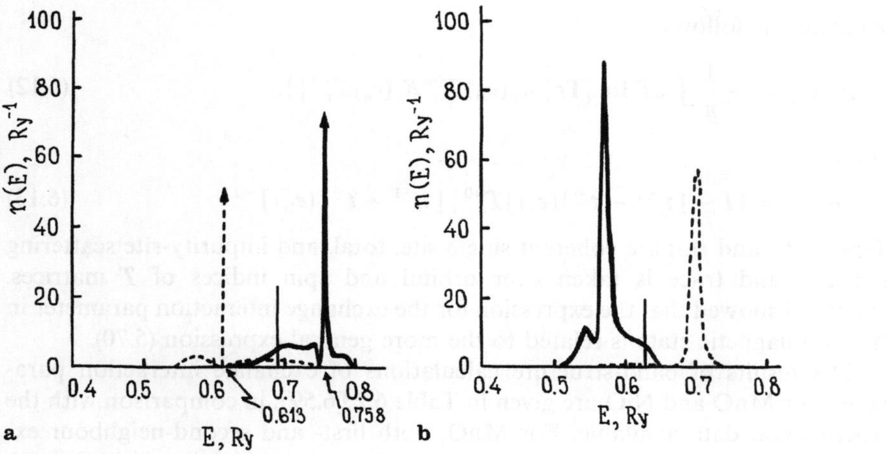

Fig. 6.24. Partial d DOS of the transition-metal-atom t_{2g} orbitals of the minority-spin state with population imbalance only at a given site in (**a**) FeO and (**b**) CoO. The broken curve is a contribution from one of the t_{2g} orbitals, which is almost empty, and the solid curve is for the other two orbitals. For FeO, the arrow at 0.613 Ry denotes a split-off state in the gap

gap for the bulk crystal. We believe that band-structure calculations including the correlation effects for the 3d elements would give a correct description for an antiferromagnetic insulator. This can be done, for example, on the basis of the prescription by *Perdew* and *Zunger* [6.67] (see also Sect. 6.1) to introduce an anisotropic orbital-dependent self-interaction correction (SIC) to the potential within the LSDF formalism.

A recent LSDF–SIC calculation [6.68] gave good results for the values of band gap and magnetic moments in 3d oxides.

Let us turn now to the calculation of superexchange interactions in transition-metal oxides. Chapter 5 described the method for calculating the effective exchange interactions in the LSDF band-structure formalism. In principle, it is possible to investigate the exchange interactions in the real AF–II structure of 3d oxides, which would provide results different from those obtained for ferro- or paramagnetic structures. Superexchange interactions have been calculated for the paramagnetic configuration [6.59]. The exchange interaction energy E_n^{ex} and the exchange parameter J_n can be defined as

$$E_n^{ex} = \tfrac{1}{2}[\Delta E_n(0) - \Delta E_n(\pi)],$$
$$J_n = -E_n^{ex}/S^2, \qquad (6.11)$$

where $E_n^{ex}(\theta)$ is the change of the energy of paramagnetic crystal due to the deviation of the magnetic moment associated with the atom at $\boldsymbol{R}_i$ from the direction of the magnetic moment of the reference atom at $\boldsymbol{R}_0$. It can be calculated as follows:

$$E_n^{ex}(\theta) = -\frac{1}{\pi} \int^{E_F} dE \, \mathrm{Im}\{\mathrm{Tr}\{\boldsymbol{K}_0(e_0)\boldsymbol{T}_c^{0n}\boldsymbol{K}_n(e_n)\boldsymbol{T}_c^{n0}\}\}, \tag{6.12}$$

where

$$\boldsymbol{K}_0(e) = \{\boldsymbol{1} - [\boldsymbol{t}_c^{-1} - \boldsymbol{t}^{-1}(e_n)]\boldsymbol{T}_c^{00}\}[\boldsymbol{t}_c^{-1} - \boldsymbol{t}^{-1}(e_0)]^{-1}. \tag{6.13}$$

Here $\boldsymbol{t}_c$, $\boldsymbol{T}_c$, and $\boldsymbol{t}(e)$ are coherent single-site, total, and impurity-site scattering matrices, and trace is taken over orbital and spin indices of $\boldsymbol{T}$ matrices. Chapter 4 showed that the expression for the exchange interaction parameter in the paramagnetic state is related to the more general expression (5.70).

The results of band-structure calculations of exchange interaction parameters for MnO and NiO are given in Table 6.2 [6.59], in comparison with the experimental data available. For MnO, both first- and second-neighbour exchange parameters are negative and of the same order, which is consistent with experimental estimates. At the same time, a noticeable difference between calculated and experimental values occurs for NiO. LSDF band-structure calculation overestimates the values of J_n by a factor of 6, as compared with the most accurate spin-wave measurements [6.73]. However, the qualitative conclusion that the exchange interaction between nearest neighbours is weak and ferromagnetic, while that between second-nearest neighbours is comparatively strong and antiferromagnetic, is quite consistent with experiment.

To describe the general features of the superexchange interaction, as is possible on the basis of the band-structure calculations, it is useful to plot E_n^{ex} as a function of E_F. Figure 6.25 shows the dependence of E_n^{ex} on band filling in comparison with partial t_{2g} and e_g densities of states of MnO and NiO [6.59]. The analysis of the relationship between the magnetic interactions and the d band filling shows that for a nearly half-filled d band exchange interaction between neighbouring magnetic moments is generally antiferromagnetic, whereas for an almost empty or almost filled d band, it is ferromagnetic. This is basically valid for $3d$ oxides, but the situation in actual systems is complicated

Table 6.2. Experimental and theoretical [6.59] exchange parameters of the first and the second neighbours of MnO and NiO

MnO		NiO		Reference
J_1 [K]	J_2 [K]	J_1 [K]	J_2 [K]	
−14.4	−7.0	−50	−85	[6.69]
−9.0	−10.4			[6.70]
−10.0	−11.0			[6.71]
−28	−28			[6.72]
		16	−222	[6.73]
Calculation results:				
−30.3	−29.8	61	−1230	[6.59]

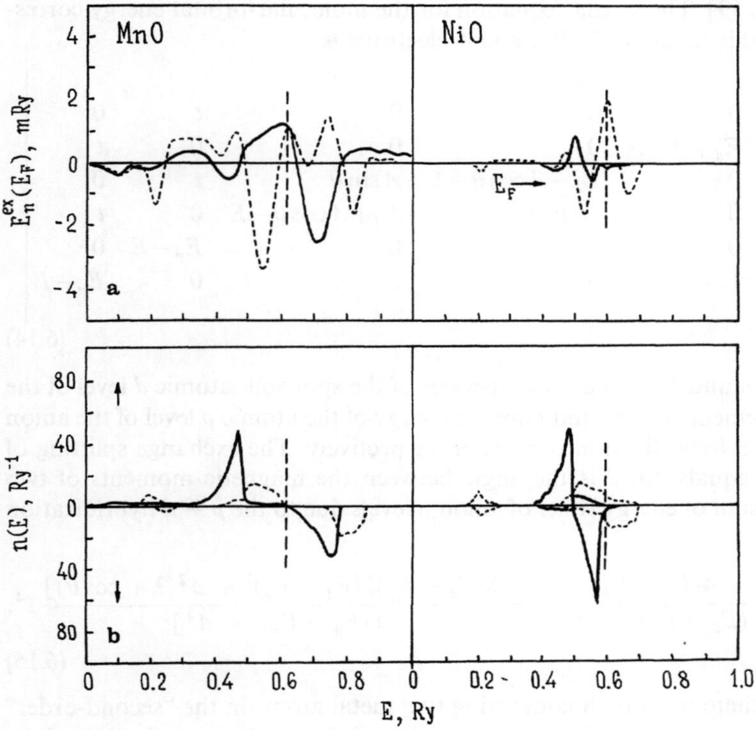

Fig. 6.25. (**a**) Exchange interaction energy E_n^{ex} for $n = 1$ (*solid line*) and $n = 2$ (*broken line*) as a function of E_F for MnO and NiO; (**b**) Partial DOS with t_{2g} (*solid line*) and e_g (*broken line*) symmetries at the Mn or Ni site

by the fact that the superexchange coupling is mediated by the *d*-band covalence effects.

In the case of MnO, where the spin-up *d* band is full and the spin-down band is empty, both first- and second-neighbours exchange parameters calculated at the actual position of the Fermi level are antiferromagnetic. Figure 6.25 also shows that both t_{2g} and e_g states contribute considerably to the exchange interaction, because E_n^{ex} strongly varies as a function of E_F in the energy regions where the t_{2g} and e_g subbands are located.

For NiO, the nearest-neighbours exchange interaction J_1 is mainly related to the t_{2g} subband, while second-nearest interaction J_2 is determined by a strong σ interaction via the e_g band. Taking into account that the t_{2g} subband is almost filled and the e_g subband is half-filled in NiO, it is easy to understand that $J_1 > 0$ and $J_2 < 0$, with $|J_1| \gg |J_2|$ in this case.

To analyse the difference between the Anderson theory of superexchange and the band-structure calculations of magnetic coupling in solids, is is useful to consider a simple model of six orthogonal orbitals imitating the *d-p-d* bond

(Fig. 6.26) [6.59]. The secular equation for the molecular-orbital energy corresponding to this model of 3 sites and 4 electrons is

$$\begin{vmatrix} E_d-\Delta-E & 0 & 0 & 0 & t & 0 \\ 0 & E_d+\Delta-E & 0 & 0 & 0 & t \\ 0 & 0 & E_d-\Delta\cos\theta-E & \Delta\sin\theta & t & 0 \\ 0 & 0 & \Delta\sin\theta & E_d+\Delta\cos\theta-E & 0 & t \\ t & 0 & t & 0 & E_p-E & 0 \\ 0 & t & 0 & t & 0 & E_p-E \end{vmatrix} = 0. \tag{6.14}$$

where $E_d - \Delta$ and $E_d + \Delta$ are the energies of the spin-split atomic d level of the transition element, and E_p and t are the energy of the atomic p level of the anion and the $p - d$ hybridization parameter, respectively. The exchange splitting of the d states equals 2Δ; θ is the angle between the magnetic moments of two atoms. The sum of energy shifts of anion p levels due to the $p - d$ hybridization is given by

$$\Delta E = -\frac{4(E_d - E_p)}{(E_d - E_p)^2 - \Delta^2} t^2 + \frac{8(E_d - E_p)[(E_d - E_p)^2 + \Delta^2(2 + \cos\theta)]}{[(E_d - E_p)^2 - \Delta^2]^3} t^4. \tag{6.15}$$

As soon as there is no path connecting two metal atoms in the "second-order" perturbation, the t^2 term does not depend on θ. Then, the contribution of the anion p state to the superexchange energy is equal to

$$E^{xc}_{\text{anion}} = \tfrac{1}{2}[\Delta E(\theta = 0) - \Delta E(\theta = \pi)] = \frac{8(E_d - E_p)\Delta^2}{[(E_d - E_p)^2 - \Delta^2]^3} t^4. \tag{6.16}$$

This contribution is antiferromagnetic and quite sensitive to the values of exchange splitting of d states Δ and the "charge transfer" energy $(E_d - E_p)$. In the Anderson superexchange model, the magnetic coupling is proportional to the inverse value of the Coulomb repulsion parameter U, while in the LSDF-based band structure, the Hund's rule energy Δ takes its place.

Figure 6.27 shows the general trends in the T_N change over the row of $3d$ oxides. All calculated values are normalized to the experimental value for NiO

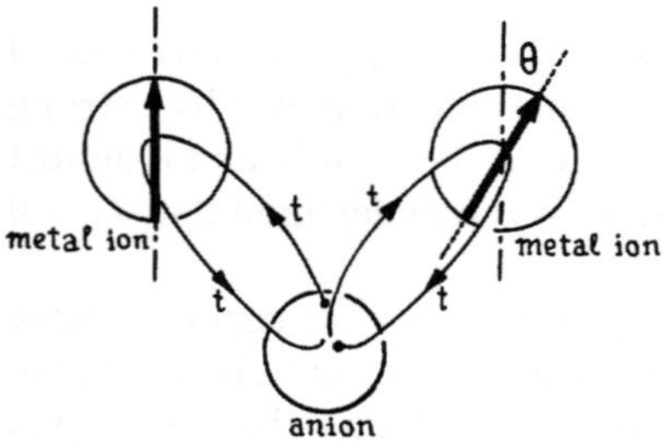

Fig. 6.26. Six-orthogonal-orbital model of d–p–d hybridization

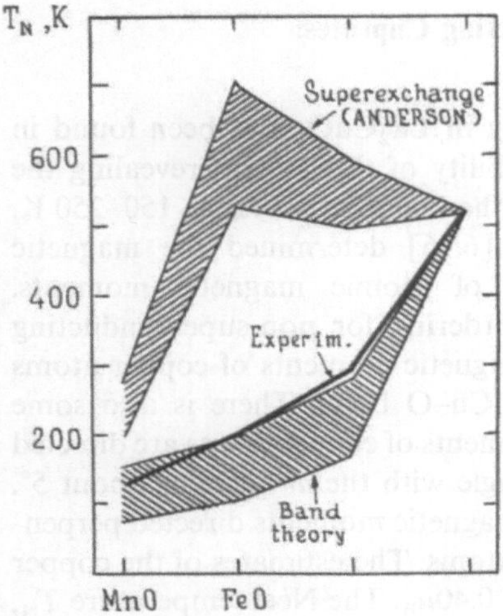

Fig. 6.27. Néel temperature of the monoxides according to the superexchange model, band-structure calculations and experimental data [6.62]

[6.62]. The overall tendency can be easily understood: at the end of the transition-metal series, the d shell is filled, the atomic magnetic moment decreases and exchange splitting Δ is reduced. Since Δ replaces U as the "energy denominator" and since the hopping t is approximately constant, the Néel temperature rises.

The main problem in the LSDF band-structure calculation is that the decrease in Δ when going from MnO to NiO is strongly overestimated, and hence the antiferromagnetic exchange interaction parameter calculated for NiO is somehow too high. It is possible that the introduction of the self-interaction correction or other non-spherical correction to the exchange-correlation potential taken in the local approximation would result in bigger spin splitting within the d band, and therefore reduce the estimated Néel temperature for NiO, FeO and CoO to more realistic values.

6.4 Magnetic Structure and Exchange Interactions in High-Temperature Superconductors

Distinct peculiarities of electronic structure and chemical bonding in high-temperature superconductors are closely related to their unusual magnetic characteristics. A number of theoretical models of superconductivity developed up to now [6.74, 75] include magnetic excitations as a principal element of superconductivity mechanisms. This is the reason for the lively interest in and the large number of experimental and theoretical papers devoted to the magnetic structure of high-T_C materials.

6.4.1 Magnetic Ordering in Non-superconducting Cuprates: Experimental Results

The possibility of definite magnetic ordering in La_2CuO_4 has been found in early experiments on the magnetic susceptibility of this system revealing the possibility of a magnetic phase transition in the temperature range 150–250 K. Later on, neutron scattering experiments [6.76] determined the magnetic ordering type and evaluated the values of atomic magnetic moments. Figure 6.28a shows the type of magnetic ordering for non-superconducting La_2CuO_4. Antiferromagnetically ordered magnetic moments of copper atoms form a 45° angle with the direction of the Cu–O bond. There is also some experimental evidence that spin magnetic moments of copper atoms are directed together with CuO_6 octahedra, and their angle with the *ab* plane is about 5°. Oxygen atoms O_{xy} also possess some small magnetic moments directed perpendicular to the magnetic moments of copper atoms. The estimates of the copper magnetic moment μ_{Cu} lie between $0.23\mu_B$ and $0.40\mu_B$. The Néel temperature T_N, according to magnetic susceptibility data, is in the interval from 150 to 290 K, and depends strongly on the oxygen nonstoichiometry. In systems doped with metals with valence 2 (Ba, Sr), it depends on the concentration of these metals. The antiferromagnetic ordering disappears quickly with the increase of metal concentration. Nevertheless, up to $x = 0.06$ in the $La_{2-x}(Ba, Sr)_xCuO_4$ system, antiferromagnetic ordering in the copper-oxygen planes remains unchanged [6.77, 78]. In pure La_2CuO_4 at temperatures higher than T_N, characteristic correlation lengths of magnetic interactions have an order of magnetic moment in copper atoms.

Similar to the superconducting solid solutions $La_{2-x}M_xCuO_4$, superconducting compositions of the $YBa_2Cu_3O_{7-\delta}$ system do not show any magnetic ordering. However, for non-superconducting compositions, and first of all for $YBa_2Cu_3O_6$, a magnetically ordered state of the system is distinctly observed. Figure 6.28b shows the magnetic ordering type for $YBa_2Cu_3O_6$. Magnetic moments of copper atoms in the planes are ordered antiferromagnetically. The

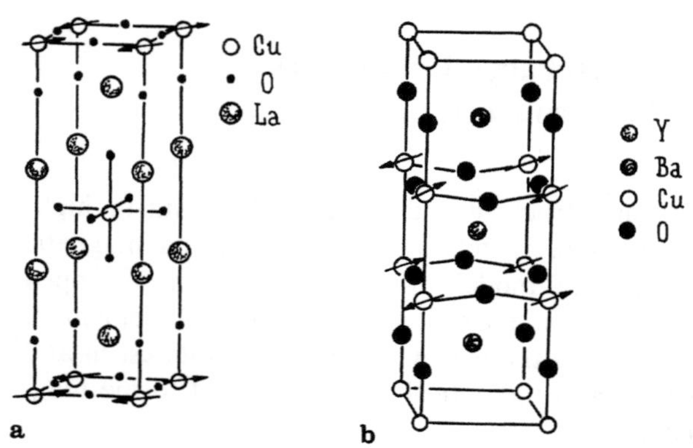

Fig. 6.28. Magnetic structure of non-superconducting La_2CuO_4 (**a**) and $YBa_2Cu_3O_6$ (**b**)

Néel temperature for the given ordering type is strongly dependent on δ and equals 500 K for $\delta = 0.0$, 450 K for $\delta = 0.1$, 230 K and for $\delta = 0.35$. The values of Cu2 magnetic moments are $0.97\mu_B$ (estimates [6.79, 80] give $\mu_{Cu2} = 0.5\mu_B$). Magnetic ordering of Cu1 ions appears at temperatures below $T_{N_2} = 80$ K for $\delta = 0.1$, and $T_{N_2} = 10$ K for $\delta = 0.35$. The estimate of the Cu1 magnetic moment is $0.46\mu_B$. Some other estimates of Néel temperatures T_{N_1} for $0.0 < \delta < 0.4$ have been reported in [6.81] and give 450 K for $\delta = 0.0$. Magnetic ordering disappears already for $\delta = 0.3$–0.35 [6.81]. The values of T_{N_1} and T_{N_2} for monocrystal samples with $\delta = 0.0$ and 0.35 are 405 K and 40 K respectively [6.82], and the value of the Cu1 magnetic moment is perhaps significantly underestimated: $\mu_{Cu1} = 0.02\mu_B$.

Thus, there is some inconsistency in the values of magnetic ordering temperatures and magnetic moments of constituent atoms, especially those obtained with different experimental methods [6.81, 82]. Nevertheless, the type of magnetic ordering in the compounds which form the basis for high-T_C superconductors is quite obvious, as is the order of magnitude of the most characteristic to be far from trivial, and up to now is not completely satisfactory.

6.4.2 Band-Structure and Cluster Calculations of Antiferromagnetic Ordering

The main question which should be answered by electronic structure calculations of magnetic crystals concerns the nature of the most stable spin configuration of the crystal. That is why just after spin-restricted nonmagnetic calculations of electronic structure and chemical bonding of high-T_C superconductors, several attempts have been undertaken to reproduce in the calculations (mainly in the scope of the local spin-density approximation) finer effects connected with small energy differences of the states with different orientations of magnetic moments. It is not so easy to calculate these differences in the scope of the usual LSDF band-structure formalism, and different authors' results are often contradictory.

Leung et al. [6.83] were not able to obtain the experimentally observed antiferromagnetic ordering in La_2CuO_4. The calculated results of *Shiraishi* et al. [6.84] led to an antiferromagnetic ground state of this compound with the antiferromagnetic gap value $\Delta \sim 0.6$ eV (Fig. 6.29) and the copper magnetic-moment value $\mu = 0.43\mu_B$. *Hatsugai* and *Fujiwara* [6.85] have found an antiferromagnetic solution for La_2CuO_4 in the scope of the LMTO method with the exchange-correlation potential used by von Barth and Hedin. The magnetic-moment value appeared to be $0.15\mu_B$ per copper atom, and the energy of the antiferromagnetic configuration was 50 mRy lower than the energy of the paramagnetic state. Similar results have been obtained elsewhere [6.88]; however, the energy difference of the paramagnetic and antiferromagnetic phases was strongly underestimated ($\sim$ 1 mRy), and the value of the copper magnetic moment depended strongly on the details of the calculations. Similar strong dependence of calculation results on the shape of the exchange-correlation potential has been found in APW band-structure cauculations [6.87], where,

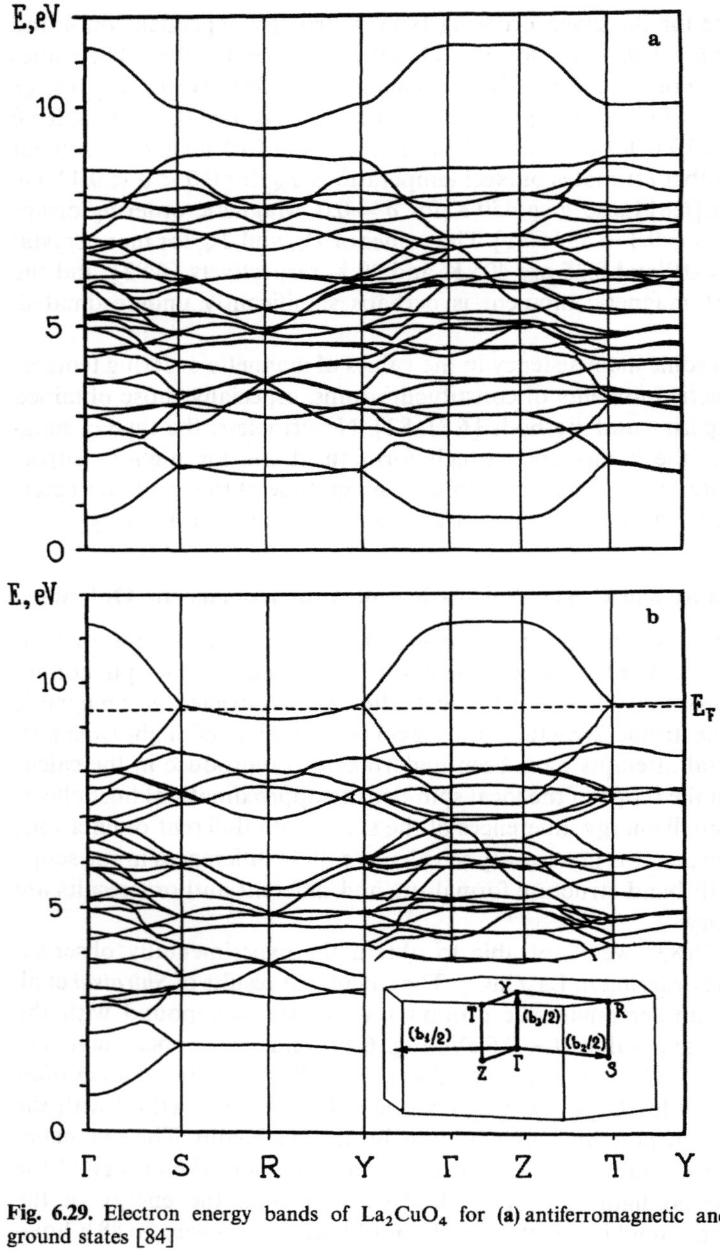

Fig. 6.29. Electron energy bands of La_2CuO_4 for (a) antiferromagnetic and (b) non-magnetic ground states [84]

depending on the type of potential, either nonmagnetic or antiferromagnetic solutions (with $\mu \simeq 0.4\mu_B$ and $\Delta \sim 0.1$ eV) were obtained.

Using the X_α discrete variational approach, some cluster modelling of the magnetic order in La_2CuO_4 has been carried out [6.88]. The results led to the conclusion that calculation results may depend strongly not only on the details of the calculational methods used but also on the crystal lattice distortions. In fact, spin-polarized calculations of the $Cu_2O_7^{10-}$ clusters with bond lengths corresponding to experimental lattice parameters of La_2CuO_4 ($R_{Cu-O} = 1.9$ Å) lead to antiferromagnetic ordering of magnetic moments. This state is the only stable magnetic state of the cluster. The value of the magnetic moment in the copper atom is $0.24\mu_B$; the difference in total energies of magnetic and non-magnetic configurations equals -0.01 eV. However, even for small distortions of the crystal lattice, for example, when the distance between the Cu and oxygen atoms decreases by only 3%, the non-magnetic solution appears to be the more stable, and the magnetic moment at the copper atom disappears completely. Independently of the Cu–O distances, any attempts to get ferromagnetic ordering of magnetic moments on the cluster result in a non-magnetic solution.

The case of $YBa_2Cu_3O_6$ is even unclearer. *Szpunar* et al. obtained an antiferromagnetic solution of the band-structure problem for $YBa_2Cu_3O_6$ as the most stable state, making use of the LMTO method [6.89]. But band structures calculated by the same method for superconducting $YBa_2Cu_3O_7$ differ essentially from the results of other authors obtained both with the LMTO and other methods. The values of copper magnetic moments are $\mu_{Cu1} \approx 0$, $\mu_{Cu2} = 0.38\mu_B$; the value of the antiferromagnetic gap at E_F equals 0.1 eV [6.89]. Another calculation [6.90] by the same LMTO method did not lead to so definite conclusions: depending on the calculation parameters (the number of $\boldsymbol{k}$ points in the Brillouin zone, the number of basis orbitals), either a nonmagnetic or an antiferromagnetic solution with a small value for the magnetic moment at the copper atom was obtained [6.90]. Band-structure calculations in the scope of local spin-density formalism are not sufficient to obtain definite results on copper magnetic-moment values corresponding to the experimental picture.

Thus, for both La_2CuO_4 and $YBa_2Cu_3O_6$, the results of different band-structure calculations do not agree on the most stable magnetic configuration of the system. But they are rather consistent on the value of magnetic moments: all calculations definitely underestimate the magnetic moments at the copper atoms and the value of the antiferromagnetic gap in the case when the calculations give the antiferromagnetic configuration as the ground state. The conclusion most appropriate at the moment is that along with the "numerical" defficulties in obtaining exact solutions of the band-structure problem in the scope of the local spin-density approximation, the principal difficulty is the proper account of correlation effects, which are very important for the systems considered. Different approaches have been suggested which in principle overcome the drawbacks of the local density approximation and directly take into account correlation effects. However, most still contain many not very well defined parameters [6.91–94] or parameters estimated from the calculations of small clusters [6.95, 96]. Probably, non-empirical calculations of the antiferromagnetic state in cuprates could be feasible when taking into account the self-interaction corrections [6.68].

However, there is another approach which allows the detailed study of magnetic interactions in both non-superconducting and superconducting cuprates, the approach based on the band-structure calculations of non-magnetic states. This approach [6.97] can be applied to the high $- T_C$ superconductors with not too large magnetic moments at copper atoms.

In second order perturbation theory, the change of total energy due to magnetic moments appearing at atoms of an initially non-magnetic configuration is given by

$$\Delta E^{(2)} = -\frac{1}{4}\sum_{ij}\sum_{LL'} I_i I_j \chi^{ij}_{LL'} m_i m_j. \tag{6.17}$$

Here $\chi^{ij}_{LL'}$ is the non-local magnetic susceptibility of atoms i and j (see Sect. 3.5):

$$\chi^{ij}_{LL'} = \frac{1}{\pi}\,\mathrm{Im}\left\{\int^{E_F} dE\, G^{ij}_{LL'}(E) G^{ji}_{L'L}(E)\right\}. \tag{6.18}$$

As is obvious from (6.17), the type of magnetic ordering is reflected already in the sign of $\chi^{ij}_{LL'}$: the antiferromagnetic ordering corresponds to negative χ_{12} values; ferromagnetic ordering gives positive values of χ_{12}.

The exchange interaction parameters defined from the results of spin-restricted band-structure calculations are given then as

$$J_{ij} = 2 I_i I_j \sum_{LL'} \chi^{ij}_{LL'}. \tag{6.19}$$

Note that for copper oxide superconductors this relation allows one to evaluate, for example, superexchange interaction Cu–O–Cu parameters, including both direct and indirect exchange through oxygen O 2p states.

6.4.3 Exchange Interaction Parameters in High-T_C Superconductors

As has been shown above, the stability of the antiferromagnetic state in La–Sr–Cu–O and Y–Ba–Cu–O systems, as well as the character of magnetic Cu–O–Cu interactions, is closely connected with the nature of high-T_C superconductivity. Detailed information on the magnetic characteristics of the crystals can be obtained from calculations on non-local magnetic susceptibilities (6.18). We have carried out a number of such calculations for high-T_C systems.

Figure 6.30a shows the dependence of non-local magnetic susceptibility on the Fermi energy for a pair of copper atoms, connected by the vector $\boldsymbol{R}_{ij} = (0\ 1\ 0)$, in La_2CuO_4. As is seen, χ_{dd} depends strongly on the Fermi energy, but in a rather wide interval around the actual position of E_F magnetic interactions are essentially antiferromagnetic. Dependence of χ_{dd} on E_F for the pair of copper atoms from the CuO_2 planes in $YBa_2Cu_3O_7$ (Fig. 6.30b) is quite similar, and it also has an antiferromagnetic character near the Fermi level. But the shape of $\chi_{dd}(E_F)$ for the pair of copper atoms belonging to the chains in $YBa_2Cu_3O_7$ (Fig. 6.30c) essentially differs, and near the Fermi level it is of a ferromagnetic character.

Table 6.3 gives some results of the non-local magnetic susceptibility calculations and exchange interaction parameters for different pairs of atoms in

Table 6.3. Calculated exchange interaction parameters in $YBa_2Cu_3O_7$ and La_2CuO_4 (for $S = 1/2$ and $I_{Cu} = 0.07$ Ry, $I_O = 0.11$ Ry)

Compound	Bond	R_{ij}	χ^{ij}_{dd} [mRy]	J_{ij} [K]
$YBa_2Cu_3O_7$	Cu2–O2–Cu2	(010)	– 50.8	– 157
	Cu2–O3–Cu2	(100)	– 42.8	– 132
	Cu1–O4–Cu2	(001)	– 9.9	– 31
	Cu1–O1–Cu1	(010)	72.5	225
	Cu2–O2	(1/2 0 0)	112.3	545
	O2–O2	(1/2 1/2 0)	35.5	270
La_2CuO_4	Cu–O1–Cu	(100)	– 73.5	– 227
	Cu–O2–Cu	(001)	– 0.45	1.4

La_2CuO_4 and $YBa_2Cu_3O_7$. For the layered perovskite La_2CuO_4, strong anisotropy of exchange interactions appears in the *ab* plane and along the *c* axis. Estimates of antiferromagnetic ordering temperature, which can be made using (6.19), agree well with experimental results.

The magnetic structure of the superconductor $YBa_2Cu_3O_7$ is much more complicated (Table 6.3). There is a strong antiferromagnetic Cu–O–Cu interaction in the CuO_2 plane, which leads to the antiferromagnetic ground state in the defect-structure $YBa_2Cu_3O_6$. Superexchange of the Cu1–O4–Cu2 type has an antiferromagnetic character also, which is consistent with experimental data on antiferromagnetic ordering and the T_{N_2} value. Using the simplest mean-field approximation

$$T_N = \frac{|J|zS(S+1)}{3k}, \tag{6.20}$$

where z is the number of the nearest magnetic atoms, one may estimate the Néel temperature from the data listed in Table 6.3. These values are $T_{N_1} = 265 - 314$ K, $T_{N_2} = 61$ K. The agreement with experimental values [6.79, 98] is rather encouraging. It is important that the large ferromagnetic exchange in Cu1–04–Cu1 chains (which is experimentally confirmed in systems with $x = 0.3$–0.4 [6.99–101]) does not influence the antiferromagnetic ordering, as for $x = 1$ all chains are broken.

Rather strong ferromagnetic Cu–O and O–O interactions within the CuO_2 planes of the $YBa_2Cu_3O_7$ system seem quite unusual. This makes sense in the presence of unpaired holes in the oxygen band only. So, these calculations can be considered as a confirmation of the magnetic pairing mechanism of the oxygen holes in high-T_C superconductors [6.96].

The possibility of small magnetic moments appearing at oxygen atoms is revealed by results [6.102], where in calculations of copper–oxygen planes show that the polarized magnetic copper ion can be surrounded by oxygen ions with non-vanishing spin moments, two of which have "spin-up" and two "spin-down" directions. Experimental measurements of magnetic form factors [6.103]

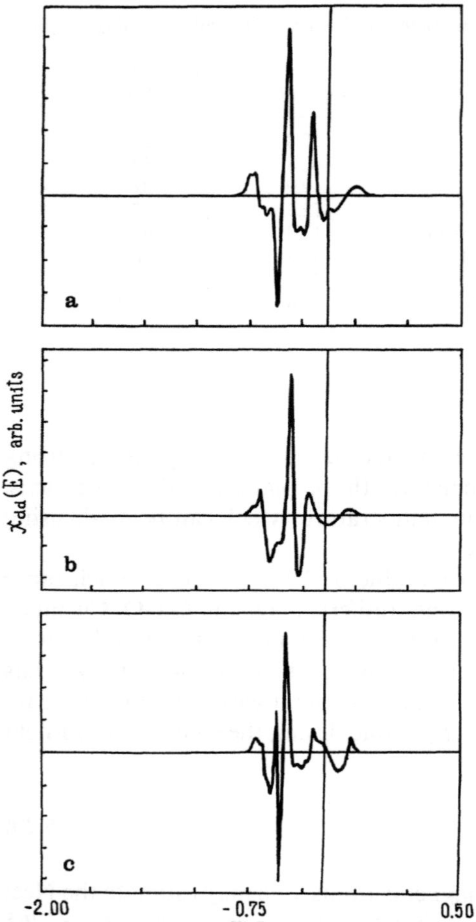

Fig. 6.30. Non-local magnetic susceptibility $\chi_{dd}(E_F)$ in La_2CuO_4 (**a**); (**b**) for Cu atoms in planes and (**c**) in chains of $YBa_2Cu_3O_7$

do not exclude the possibility of small magnetic moments existing at oxygen atoms as well.

Thus, exchange parameter calculations show that because of the strong covalence effects, mainly due to the hybridization of Cu 3*d* and O 2*p* states, magnetic interactions in high-T_C superconductors appear to be rather complicated. Magnetic bonding exists both in copper–oxygen planes, where it results in a large antiferromagnetic superexchange of the Cu2–02–Cu2 type, and between copper atoms within chains. For the latter case, the magnetic interactions are ferromagnetic. At the same time, holes in the oxygen band lead to the appearance of non-trivial magnetic bonds between copper and oxygen atoms and among oxygen atoms. Their role in high-T_C superconductivity mechanisms is not clear yet, and will be the subject of further studies.

7. Magnetic Impurities in Metals

In the previous chapter we considered some magnetic crystals, which possess perfect three-dimensional periodicity. But no less important are systems (dilute magnetic alloys etc.) in which magnetic atoms are dissolved in non-magnetic matrices, or magnetic atoms of one type are embedded in crystals formed by other magnetic atoms. The electronic structure of magnetic impurities is discussed in this chapter.

7.1 Impurities in Aluminium

Aluminium is believed to be a typical example of a simple metal, its electronic structure being described well with the free-electron approximation [7.1]. Therefore, one can expect the electronic structure of $3d$ impurities in Al to be consistent with the virtual bound state (VBS) model proposed by *Friedel* and *Anderson* [7.2, 3]. This means that the local density of states (LDOS) on an impurity site resembles a resonance-type Lorentzian curve:

$$n(E) = \frac{5}{\pi} \frac{\Delta}{(E - E_d)^2 + \Delta^2} . \tag{7.1}$$

Here the half-width parameter $\Delta = (2\pi/\hbar)/|V_{sd}|^2 n_{\mathrm{Al}}(E_d)$ is determined by the averaged s–d hybridization matrix element V_{sd} and by the density of states of the host metal at the energy of the impurity d level.

Experimental detection of VBS-like peaks in the valence-band X-ray photoelectron spectra of Al-based dilute alloys [7.4, 5] gave rise to theoretical efforts to study the electronic structure of these systems with numerical calculations. The electronic structure and some measurable physical properties of $3d$ impurities in Al had been calculated even earlier [7.6–8] using the simplest "jellium" model. Then a self-consistent calculation was performed [7.9] in the framework of a more sophisticated "spherical solid" model. In this model, a single muffin-tin sphere includes not only the impurity atom but several neighbouring atomic shells as well. The potentials of the atoms in these shells are averaged over angles with respect to the impurity position. The effects of "backscattering" of electrons by neighbouring atoms are thus, although very approximately, taken into account in the calculation. This considerably improved agreement with the experimental data on the widths and positions of impurity virtual bound states (Table 7.1).

Figure 7.1b shows the calculated LDOS of $3d$ impurities in Al [7.9]. As the energy of the impurity $3d$ level is reduced on going from Sc to Zn, the corresponding VBS become narrower, as one should expect from (7.1). For Mn, the energy of the impurity $3d$ state falls in the vicinity of the Fermi level, and the VBS width is at the same time small enough to satisfy the Anderson criterion for the existence of a local magnetic moment. As follows from estimates [7.9], the Mn impurity in Al has a local magnetic moment of $2.46\mu_B$, which is consistent with experimental evidence [7.10] that AlMn is a spin-fluctuating system.

In the simplest numerical model for impurities in Al–the "jellium" model–it is possible to describe only qualitatively how the residual resistivity $\Delta\rho$ changes in the sequence of $3d$ impurities. Figure 7.2 shows results of calculation for $\Delta\rho$ [7.11] in comparison with the experimental data. Even when only one-site scattering is taken into account (i.e. in the jellium model), the calculation still reproduces the overall shape of the $\Delta\rho(Z)$ curve, with a steep maximum in the middle of the $3d$ row (which reveals the enhancement scattering of conduction electrons on VBS of Cr and Mn which actually cross the Fermi level). Approximate inclusion of multiple-scattering effects in the calculation [7.11] improved the agreement with experimental results on the detailed tendency of $\Delta\rho$ changing through the $3d$ row. Numerical agreement of numerical values found for different impurities was, however, still far from excellent.

Other examples of calculated physical properties such as electronic specific heat and thermoelectric power for Al-based dilute alloys can be found in the literature [7.6–8].

Table 7.1. Characteristic parameters of virtual bound states for $3d$ impurities in Al according to experimental data and calculation results

	Impurities							Method of research
	V	Cr	Mn	Fe	Co	Ni	Cu	
E_d	—	—	−0.72	−0.93	−1.53	−2.42	−4.52	X-ray photo-
Δ	—	—	0.72	0.77	0.82	0.61	0.56	electron spec-
n_d	—	—	7.56	7.87	8.44	9.22	9.61	troscopy [7.5]
E_d	0.17	−0.25	−0.60	−1.03	−1.84	−3.43	−6.57	"Impurity
Δ	1.63	1.28	1.03	0.82	0.63	0.40	0.15	in jellium" [7.8]
E_d	0.13	−0.40	−0.60	−0.82	−1.20	−1.82	−3.40	"Spherical
Δ	1.45	1.05	0.90	0.75	0.60	0.46	0.32	solid" [7.9]
E_d	0.0	−0.3	−0.6	−0.8	−1.1	−1.8	−3.1	KKR-Green
Δ	3.4	2.9	2.3	2.0	1.5	0.6	0.7	function method [7.12]
E_d	—	—	−0.48	−0.82	—	—	—	Cluster Green
Δ	—	—	1.36	0.76	—	—	—	function
n_d	—	—	5.65	7.50	—	—	—	method

E_d is the energy position of the VBS (in eV) with respect to the Fermi level;
Δ is the VBS halfwidth (in eV); n_d is the number of electrons in the VBS

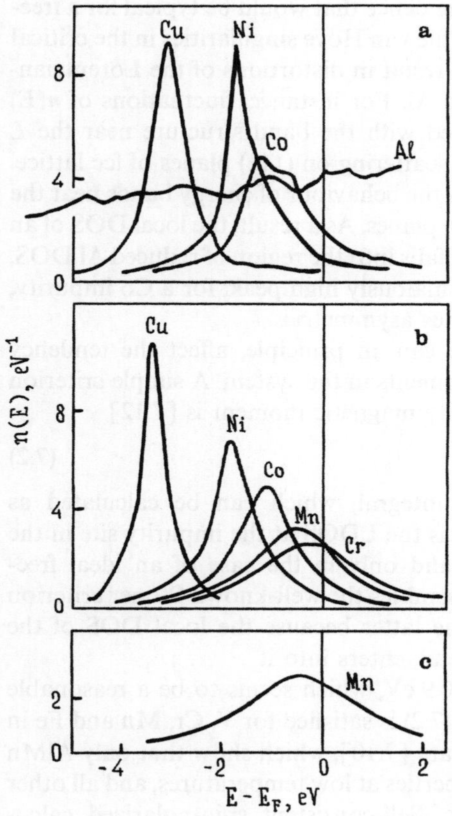

Fig. 7.1. Local DOS of $3d$ impurities in Al as calculated by **(a)** the KKR–Green function method [7.12], **(b)** "spherical solid" method [7.9] and **(c)** cluster Green function method

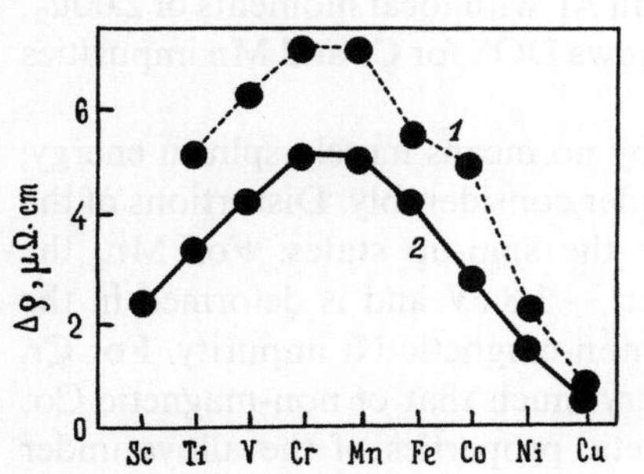

Fig. 7.2. Experimentally determined (*1*) and calculated (*2*) values of residual resistivity due to $3d$ impurities in Al [7.11]

More sophisticated calculations of the electronic structure of impurities in Al have been performed using the KKR–Green function method [7.12–13] and the LMTO-Green function method [7.14]. Local DOS of non-magnetic impurities in Al [7.12] are shown in Fig. 7.1a along with the DOS of the pure host crystal.

Deviations of $n(E)$ from the $\sim \sqrt{E}$ dependence that would be typical for a free-electron gas (these deviations arise from the van Hove singularities in the critical edge points of the Brillouin zone [7.1]) result in distortions of the Lorentzian-shaped VBS typical for $3d$ impurities in Al. For instance, fluctuations of $n(E)$ near -4.5 eV in Fig. 7.1a are connected with the band structure near the L point and are determined by the Bragg scattering on (111) planes of fcc lattice. The feature seen at $E \simeq -1.8$ eV reveals the behaviour of energy bands near the X point and hence the scattering on (100) planes. As a result, the local DOS of an Ni impurity, the $3d$ energy level of which falls into the region of reduced Al DOS, acquires the shape of a narrow and anomalously high peak; for a Co impurity, the local DOS is broadened and becomes asymmetric.

Such distortions of the VBS shape can, in principle, affect the tendency towards formation of local magnetic moments in the system. A simple criterion proposed for the existence of an impurity magnetic moment is [7.12]

$$I n_{\mathrm{loc}}(E_{\mathrm{F}}) \geqslant 1. \qquad (7.2)$$

Here I is the intra-atomic exchange integral, which can be calculated as described by *Gunnarsson* [7.15]; $n_{\mathrm{loc}}(E)$ is the LDOS at the impurity site in the non-magnetic case. This criterion is valid only in the case of an ideal free-electron-like host matrix. It formally resembles the well-known Stoner criterion for ferromagnetism, but differs from the latter because the local DOS of the impurity, and not total DOS of the metal, enters into it.

As has been shown [7.12], for $I \simeq 0.9$ eV, which seems to be a reasonable estimate for $3d$ impurities, the criterion (7.2) is satisfied for V, Cr, Mn and Fe in Al. This contradicts the experimental data [7.10], which show that only AlMn and AlCr exhibit spin-fluctuational properties at low temperatures, and all other impurities are definitely non-magnetic. Self-consistent spin-polarized calculations [7.12] have shown that V impurity does not possess a local magnetic moment, while Cr, Mn and Fe are magnetic in Al, with local moments of $2.00\mu_{\mathrm{B}}$, $2.53\mu_{\mathrm{B}}$ and $1.78\mu_{\mathrm{B}}$, respectively. Figure 7.3 shows DOS for Cr and Mn impurities [7.12].

The LDOS for two spin directions are by no means merely split in energy; the shapes of spin-up and spin-down VBS differ considerably. Distortions of the resonance peak are more pronounced for the spin-up states. For Mn, the corresponding spin-up peak is situated near -1.8 eV and is deformed in the same way as discussed above for VBS of a non-magnetic Ni impurity. For Cr, asymmetric spin-up resonance resembles very much that of non-magnetic Co.

The difficulty in interpreting the magnetic properties of the alloys under discussion is that they are spin-fluctuating systems with strong s–d interaction. Various calculations performed so far do not treat the problem of the stability of magnetic solutions found in the calculations. The only additional information extracted are the estimates of total magnetic moment in the infinite crystal due to introduction of a single impurity. These estimates based on Lloyd's formula give $1.81\mu_{\mathrm{B}}$ and $1.54\mu_{\mathrm{B}}$ for Cr, Mn and Fe impurities, respectively, thus revealing strong antiferromagnetic polarization around impurities in all three cases.

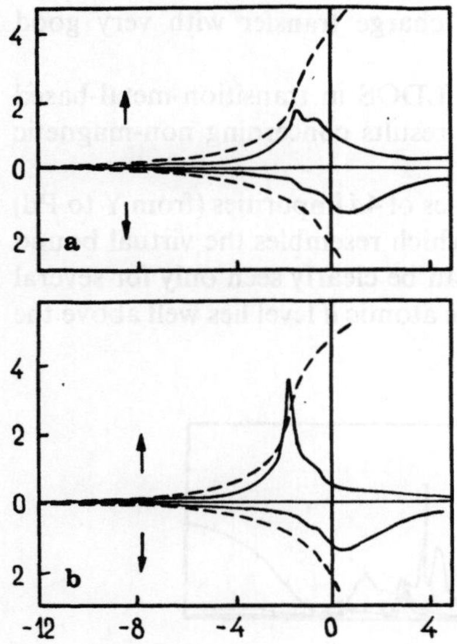

Fig. 7.3. Local DOS (*solid line*) and integrated DOS (dashed line) for magnetic impurities in Al: (**a**) Cr and (**b**) Mn

Experimental data on the values of magnetic moments of $3d$ impurities in Al are not numerous. The 2.9 eV splitting of Mn $3s$ core states found in the X-ray photoelectron spectroscopy study of AlMn alloy [7.5] made it possible to estimate the local magnetic moment at an Mn site as $2.2\mu_B$. This magnetic moment has to be stable at least on the characteristic time scale of the photoemission process, i.e. 10^{-15} s.

7.2 Impurities of Transition Metals

In the transition-metal-based impurity systems, effects of hybridization of impurity d states with the host metal d band play an important role, resulting in a rather complicated shape of the LDOS. Here simple "jellium" or "spherical solid" models are not applicable even for crude estimates. Reliable quantitative results can only be obtained based on calculational schemes which take into account anisotropic spatial distribution of atomic charge and spin densities and treat the impurity-host hybridization properly. Until the beginning of the 1980s, a variety of papers applied one or another version of the cluster approach to the impurity problem. At present, however, the more powerful KKR-Green function method has proved be feasible and gained some popularity, and many of the previous results (at least for impurities in metals) have been revised [7.16–18]. In more recent papers following the KKR- (or LMTO-) Green function formalism, potentials of host atoms in one or more spheres of neighbours around the impurity were allowed to adjust in the course of achieving self-consistency, thus enabling one to treat the charge transfer with very good accuracy [7.19].

To describe some general features of the LDOS in transition-metal-based impurity systems, we begin with a review of results concerning non-magnetic dilute alloys.

Figure 7.4 shows densities of states of a series of 4*d* impurities (from Y to Pd) in Cu [7.20]. A broad Lorentzian-like peak which resembles the virtual bound state typical for impurities in simple metals can be clearly seen only for several impurities at the beginning of the 4*d* row. Here atomic *d* level lies well above the

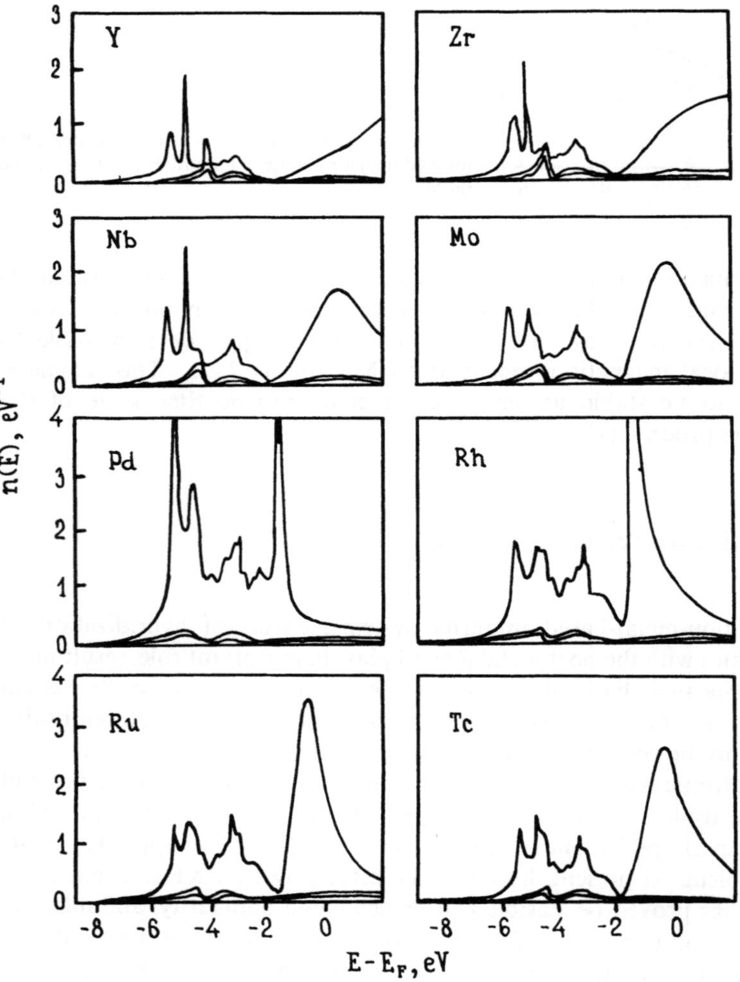

Fig. 7.4. Local DOS of 4*d* impurities in Cu [7.20]. The three lines in each plot are densities of s, $s+p$ and $s+p+d$ states

Cu *d* band so that its hybridization with the latter can be neglected. As the *d*-level energy decreases (at the end of the 4*d* series), this peak becomes more distorted and its area diminishes, whereas a complicated structure in the LDOS at lower energy, which results from hybridization with the Cu *d* band, gradually becomes more pronounced.

Following *Zeller* and *Braspenning* [7.19], we concentrate now on the CuMo alloy. Figure 7.5 shows the *d*-DOS of an Mo impurity in Cu; the result calculated using the single-site Green function without backscattering by neighbouring atoms is shown as a dashed line. Physically it illustrates the net effect of Mo 4*d* electrons' interaction with *sp* electrons of Cu, which would result in formation of a typical virtual bound state. Switching on of the *d*–*d* hybridization makes this strucure more complicated and pulls the VBS-like peak upwards from the energy region occupied by Cu *d* band. At the same time, the impurity-site DOS near the bottom of the Cu *d* band also becomes enhanced, with another resonance-like peak being formed there.

The origin of such evolution of a local impurity DOS due to its interaction with Cu *d* states can be illustrated by Fig. 7.6, which gives a scheme of hybridization. Here the *d* level of the impurity and each state from the continuum of the host metal *d* band produce a pair of bonding (low-lying) and antibonding (high-lying) states as a result of their hybridization. The energy position of corresponding levels shows that bonding states are formed mostly by host metal *d* states, whereas antibonding states have a dominating contribution from the impurity. The effect of interaction with delocalized Cu *sp* states further influences this hybridization scheme. Within the Cu *d* band where the density of *sp* states is suppressed, this interaction is weak; at the same time, the high-lying antibonding hybrid state is broadened because it interacts with the continuum of free-electron-like states. Between the "hybridized part" and the resonance peak, a dip in the local DOS is formed (it is most clearly seen for V and Cr impurities), the energy of which corresponds exactly to the top of the host metal *d* band.

The state splitting due to the effect of the crystal field plays a minor role for 3d impurities in noble metals with a fcc lattice. Actually, the main features of

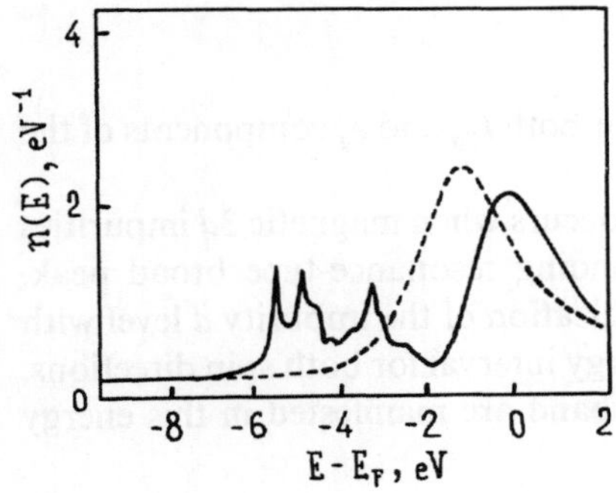

Fig. 7.5. Local DOS of Mo in Cu. *Solid line*, as calculated by the KKR–Green function method; *dashed line*, the same, with only single scattering effects by the impurity atom included in the calculation

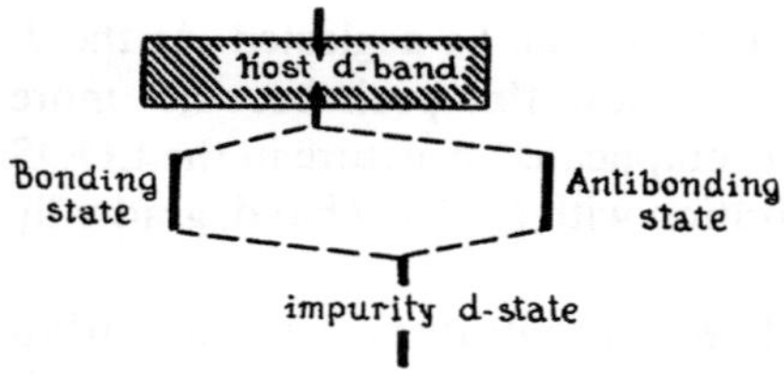

Fig. 7.6. The hybridization scheme for atomic levels of Cu and Mo

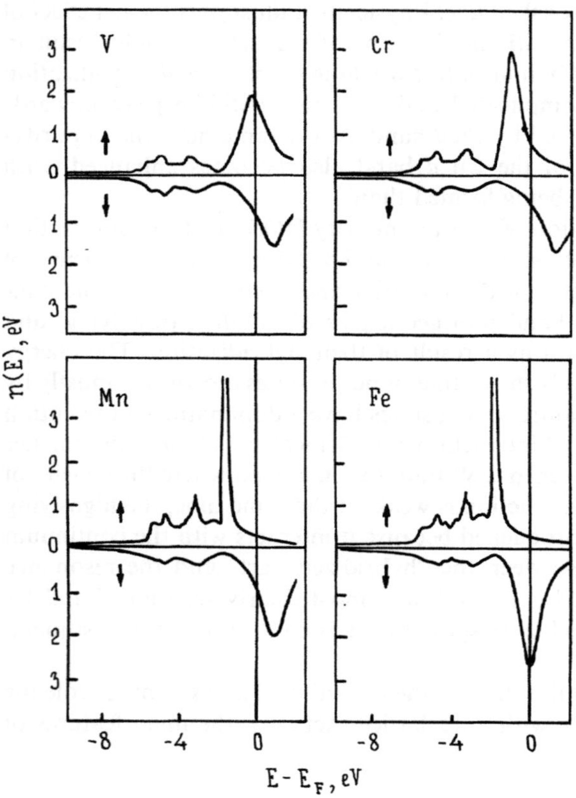

Fig. 7.7. Local DOS of magnetic impurities in Cu [7.20]

impurity DOS discussed above are typical for both t_{2g} and e_g components of the d DOS.

The spin splitting of impurity DOS that occurs when magnetic $3d$ impurities are considered affects primarily the antibonding resonance-type broad peak. The low-lying structure related to the hybridization of the impurity d level with the host d band remains within the same energy interval for both spin directions, and major peculiarities of the host metal d band are manifested in this energy region.

The most reliable calculation data concerning the electronic structure of magnetic impurities in noble metals are obtained within the KKR–Green function approach, beginning with the single-site calculations by *Zeller* and *Dederichs* et al. [7.16–18]. In subsequent papers [7.19, 20], an embedded cluster including up to four shells of neighbours to the impurity atom were introduced into the calculation. Therefore, it became possible to speak of the "solution of the embedding problem" [7.19] (within the limits set by the muffin-tin approxi-

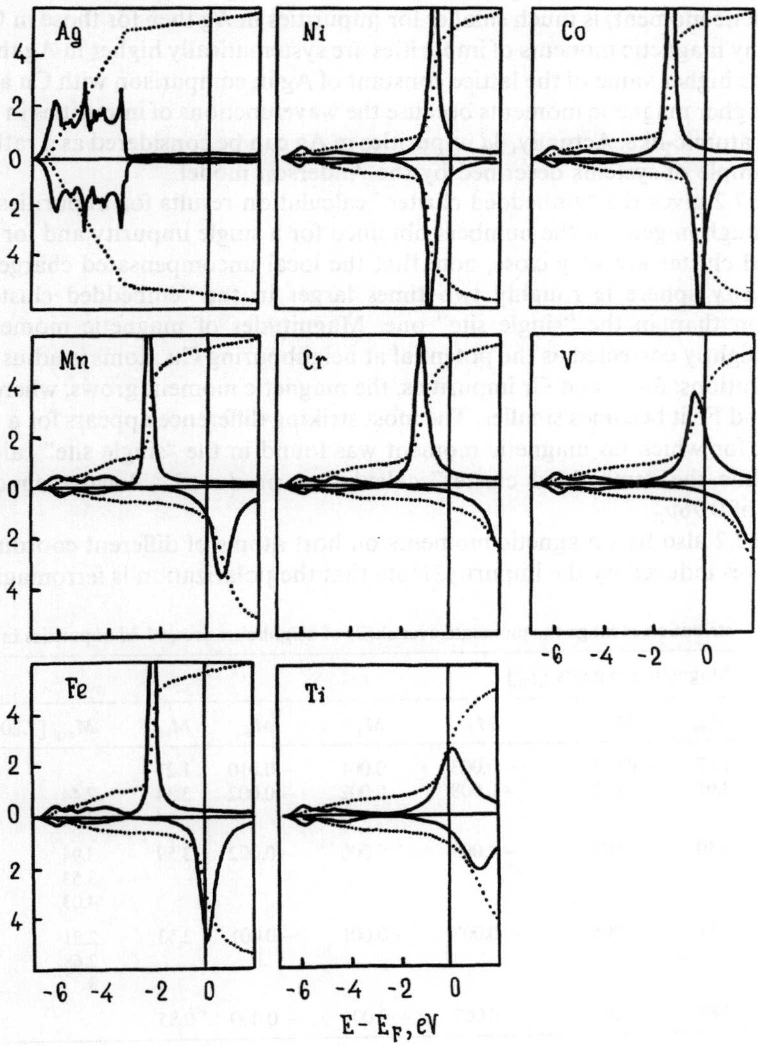

Fig. 7.8. Local DOS of pure Ag and magnetic impurities in it. The integrated DOS is shown as a dotted line

mation, local density functional theory, etc.) and to analyse the degree of localization of the impurity magnetic moment and the spatial distribution of magnetization around impurities.

Figure 7.7 shows local DOS of magnetic impurities in Cu [7.20]; Fig. 7.8 shows the DOS of pure Ag and 3*d* impurities in Ag [7.18]. As follows from the comparison of these two figures, the *d* band of Ag, which is situated lower in energy than that of Cu, exhibits rather weak hybridization with the impurity *d* level. Therefore, the low-lying "hybridization" part of the *d* DOS (which is essentially the same for both spin directions and hence gives no contribution to the magnetic moment) is much smaller for impurities in Ag then for those in Cu. This is why magnetic monents of impurities are systematically higher in Ag than in Cu. The higher value of the lattice constant of Ag in comparison with Cu also favours higher magnetic moments because the wavefunctions of impurities in Ag are more atomic-like. Actually, 3*d* impurities in Ag can be considered as a rather good example of systems described by the Anderson model.

Table 7.2 gives the "embedded cluster" calculation results for impurities in Cu. Although in general the numbers obtained for a single impurity and for an embedded cluster are very close, note that the local uncompensated charge in the impurity sphere is roughly two times larger in the "embedded cluster" calculation than in the "single site" one. Magnitudes of magnetic moments become slightly corrected as the potential at neighbouring Cu atoms is adjusted in the iterations: for V and Cr impurities, the magnetic moment grows, whereas for Mn and Fe it becomes smaller. The most striking difference appears for a Co impurity, for which no magnetic moment was found in the "single site" calculation while the "embedded cluster" calculation produced a local magnetic moment of $0.96\mu_B$.

Table 7.2 also lists magnetic moments on host atoms of different coordination spheres induced by the impurity. Note that the polarization is ferromagne-

Table 7.2. Distribution of magnetic moments over shells of neighbours around 3*d* impurities in Cu

Impurity	Magnetic moments [μ_B]						
	M_{loc}	M_1	M_2	M_3	M_4	M_{cl}	M_{exp} [7.20]
V	1.07	0.015	− 0.003	0.001	− 0.010	1.27	
Cr	3.01	0.030	− 0.008	0.002	− 0.002	3.34	2.44
							3.11
Mn	3.40	0.022	− 0.009	0.000	− 0.002	3.59	3.94
							3.53
							4.03
Fe	2.51	0.008	− 0.007	− 0.001	− 0.001	2.53	2.81
							2.68
							3
Co	0.88	0.000	− 0.002	− 0.001	− 0.000	0.85	

M_{loc} is the local magnetic moment at the impurity site;
M_n are magnetic moments of atoms within the *n*th sphere of neighbours;
M_{cl} is the total magnetic moment of the 55-atom embedded cluster

tic around all 3*d* impurities in the nearest sphere, with slight fluctuations at more distant neighbours.

7.2.1 Impurities in Pd

Group-VIII transition metals (Ni, Pd, Pt) which precede the noble metals (Cu, Ag and Au, respectively) have their filled *d* band lying just below the Fermi level. A sharp peak of the density of states near the top of the *d* band is a typical feature of fcc metals; its closeness to E_F gives rise to high susceptibility, which produces a considerable response to any magnetic fluctuation. In Ni, although the local susceptibility-based criterion for the existence of local magnetic moments (7.2) is not satisfied, the high density of states at E_F leads to the existence of itinerant magnetism, according to the Stoner criterion. Therefore, Ni-based impurities are considered in a special section below, along with the impurities in other magnetic systems. For Pd, the Stoner criterion is very close to being satisfied, and such a small amount as 0.1–0.2 at % of Fe or Co brings this metal into the ferromagnetic state.

Isolated impurities of these elements in Pd induce magnetic polarization in the host crystal with a quite large spatial extent and considerable total moment. This phenomenon is usually referred to as a giant magnetic moment. Establishing long-range ferromagnetic order in very dilute PdFe and PdCo alloys can be understood as the result of overlapping of large polarization clouds corresponding to different giant moments.

Experimental data concerning giant magnetic moments in Pd are quite numerous [7.21] and in many aspects contradictory. The magnitude of the giant magnetic moment of an Fe impurity in palladium as given by different authors varies from $4.5\mu_B$ to $12.9\mu_B$. The spatial extent of the polarization cloud also remains unclear in neutron scattering experiments. It has even been supposed [7.22] that the giant moment is localized within the first sphere of neighbours.

For other 3*d* impurities, much less experimental information is available. It remained unclear until recently whether Ni and Ti impurities in Pd possess a magnetic moment at all. It was for some time believed that a single Ni impurity remains non-magnetic, and only a cluster of three or more neighbouring Ni atoms acquires a magnetic moment [7.22, 23]. However, a thorough analysis of the experimental data [7.24] led to the conclusion that the Ni impurity in Pd is a Kondo system, with a magnetic moment of about $1\mu_B$.

Calculation of the electronic structure of 3*d* impurities in Pd within first-principles approaches was therefore very important. Of the earlier studies, the paper by *Delley* and *Ellis* [7.25] should be mentioned, where a cluster calculation by the discrete variational method was performed for the PdFe system. The total magnetic moment of the cluster of 43 atoms was found to be equal to $3.54\mu_B$, and the inclusion of one more shell of neighbours reduced the total moment to $3.24\mu_B$. The authors emphasized the importance of taking into account the magnetic polarization at Pd sites.

The KKR–Green function calculation [7.26] for an embedded cluster which included four shells of Pd atoms around a $3d$ impurity (i.e. a cluster of 43 atoms) combined the possibility of studying the spatial distribution of magnetization around the impurity with adequate treatment of band-structure effects in an infinite crystal. Table 7.3 lists calculated local magnetic moments at the impurity site and magnetic moments induced on Pd atoms in different coordination spheres.

It has been mentioned [7.26] that the values of the local magnetic moments of $3d$ impurities found in a single-site Green function calculation, i.e. with the potentials at neighbouring Pd sites kept fixed, remain practically unchanged as more shells of neighbours are included in an embedded cluster. The main effect of the cluster size concerns the magnetic polarization on Pd sites which appears and grows very slowly in the course of iterations towards self-consistency. Therefore, general features of the local DOS of impurities in Pd can be discussed on the basis of the single-site Green function calculation.

Figure 7.9 shows local densities of states of $3d$ impurities in Pd [7.27]. In most of the pictures, a typical broadened resonance level is clearly seen in the spin-down DOS well above the Pd d band. The corresponding level for spin-up states falls into the Pd d band, which lies nearer to the Fermi level than in noble metals. Therefore, the impurity d level appears to be strongly hybridized with the host d band, thus giving rise to a considerable magnitude of the impurity d DOS over the whole d band and not only at the top of the band, in contrast to the case mentioned for impurities in noble metals.

Even such a large embedded cluster as used by *Oswald* et al. [7.26] failed to include the whole space region of the polarization cloud. The three of neighbours closest to the impurity shells include less than 30% of the giant magnetic moment. Therefore, two methods were used to extrapolate to its asymptotic value [7.26].

The first method was based on an observation that the ratio of the induced magnetic moment in any coordination shell of Pd atoms to the local moment of the impurity atom in the centre obeys the same rule for all $3d$ impurities and all

Table 7.3. Distribution of magnetic moments over shells of neighbours around $3d$ impurities in Pd

Impurity	Magnetic moments [μ_B]				
	M_0	M_1	M_2	M_3	M_{cl}
Cr	3.14	− 0.024	− 0.012	− 0.004	2.69
Mn	4.13	0.048	0.011	0.015	5.12
Fe	3.47	0.102	0.028	0.028	5.53
Co	2.28	0.114	0.032	0.031	4.60
Ni	0.92	0.063	0.022	0.015	2.18

M_0 is the local magnetic moment at the impurity site;
M_n are magnetic moments of atoms within the nth sphere of neighbours;
M_{cl} is the total magnetic moment of the 43-atom embedded cluster

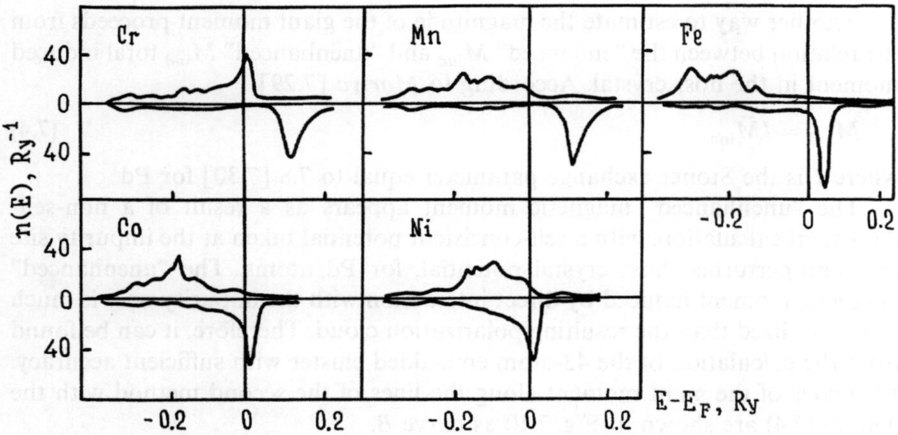

Fig. 7.9. Local DOS of $3d$ impurities in Pd

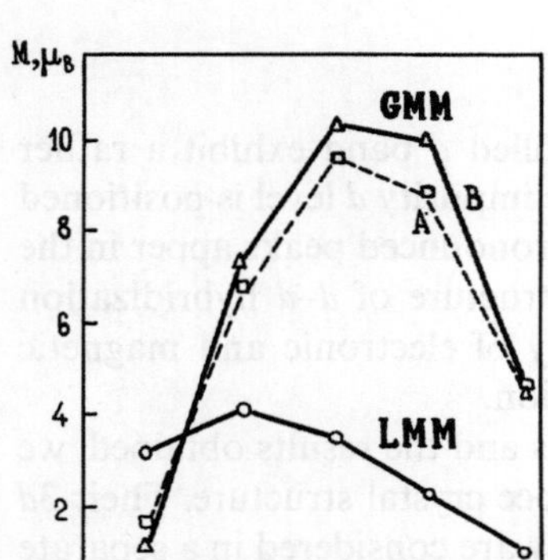

Fig. 7.10. Local magnetic moment (LMM) and giant magnetic moment (GMM) for $3d$ impurities in Pd. For GMM, estimates by methods A and B are shown (see text)

shells:

$$\frac{M_n}{M_0} \simeq \left.\frac{M_n}{M_0}\right|_{\Delta Z=0} (1 + 0.28\,\Delta Z). \tag{7.3}$$

where $n = 1, 2, 3$ runs over the shells of neighbours, and the difference in the number of valence electrons between the impurity and the host atom $\Delta Z = 0$, -1, -2, . . . for Ni, Co, Fe. Therefore, the magnitude of the giant magnetic moment for all $3d$ impurities can be estimated based on the value known for one of them. The result of this analysis with the experimental value for the giant moment of a Ni impurity ($4.6\mu_B$ [7.28]) taken as a reference point is shown in Fig. 7.10 as curve A.

Another way to estimate the magnitude of the giant moment proceeds from the relation between the "enhanced" M_{ind} and "unenhanced" $\bar{M}_{\text{ind}}$ total induced moment in the host crystal. According to *Moriya* [7.29],

$$M_{\text{ind}} = I\bar{M}_{\text{ind}}, \tag{7.4}$$

where I is the Stoner exchange parameter equal to 7.8 [7.30] for Pd.

The "unenhanced" magnetic moment appears as a result of a non-self-consistent calculation, with a self-consistent potential taken at the impurity site and non-perturbed host crystal potential, for Pd atoms. The "unenhanced" magnetic moment induced by direct interaction with the impurity only is much more localized than the resulting polarization cloud. Therefore, it can be found from the calculation of the 43-atom embedded cluster with sufficient accuracy. Estimates of the giant moment along the lines of the second method with the help of (7.4) are shown in Fig. 7.10 as curve *B*.

Both curves in Fig. 7.10, obtained by quite different methods, agree well and seem to give quite reliable results. The total moment of the polarization cloud is largest for Fe and Co and negative for Cr.

7.2.2 Impurities in Nb and Mo

Impurities in transition metals with a partially filled *d* band exhibit a rather complicated structure of the local DOS. When the impurity *d* level is positioned inside the host metal *d* band, no resonance-type pronounced peaks apper in the impurity DOS. At the same time, the detailed structure of *d–d* hybridization plays an important role in determining a variety of electronic and magnetic properties, depending on the host metal in question.

In briefly discussing the main related problems and the results obtained, we begin with 4*d* metals Nb and Mo, which have a bcc crystal structure. Their 3*d* analogs Cr and Mn are antiferromagnets, and they are considered in a separate section below. The Nb and Mo-based systems have been experimentally studied since the classical work by *Mattheiss* et al. [7.31] and *Clogston* et al. [7.32]. A challenging problem was the striking difference in the behaviour of Fe impurities in Nb and Mo: whereas an Fe atom possesses a local moment of more than $2\mu_B$ in Mo, it remains non-magnetic in Nb. Simple model approaches failed to treat on an equal footing the effects of strong hybridization of impurity states with a half-filled *d* band and a considerable crystal field splitting of t_{2g} and e_g states (by about 4 eV at the Γ point of the Brillouin zone). Note that the corresponding splitting in noble metals does not exceed the width of a virtual bound state peak.

The electronic structure of magnetic impurities in Nb and Mo was calculated by the one-site KKR–Green function method for the first time by *Podloucky* et al. [7.33]. They mentioned that the inclusion of the potential perturbation at nearest neighbours to the impurity in the iteration process would be important, because otherwise the calculation does not satisfy the Friedel sum rule. They concluded that the magnetism of Fe and Mn in Nb and Mo is covalent in character; i.e. the appearance of the magnetic moment shifts centres of gravity of two spin subbands in opposite directions, while keeping the total *d* bandwidth unchanged. In perfect crystals, this type of spin "splitting" of the local DOS is typical for antiferromagnets, as for instance Mn [7.34].

Further and more accurate study of 3*d* impurities in Nb and Mo has been done [7.35] by the KKR–Green function calculation for an embedded cluster: the impurities were almost exactly screened and the Friedel sum rule was satisfied. Figure 7.11 shows the local DOS for Fe and Mn impurities in Nb and Mo. Table 7.4 lists charges and magnetic moments per atom of impurity and its nearest neighbours according to the calculation [7.35]. Four 3*d* impurities from Cr to Co were found to possess a magnetic moment in Nb, in striking contradiction to the experimental data. Keeping in mind the quite good agreement between KKR–Green function calculations and the experimental data in evaluating magnetic moments for a number of the above-discussed impurity systems, this discrepancy seems rather unusual and needs additional discussion.

The reason for this discrepancy may be some important factor not included in the calculation, for instance, structural relaxation of the crystal around the

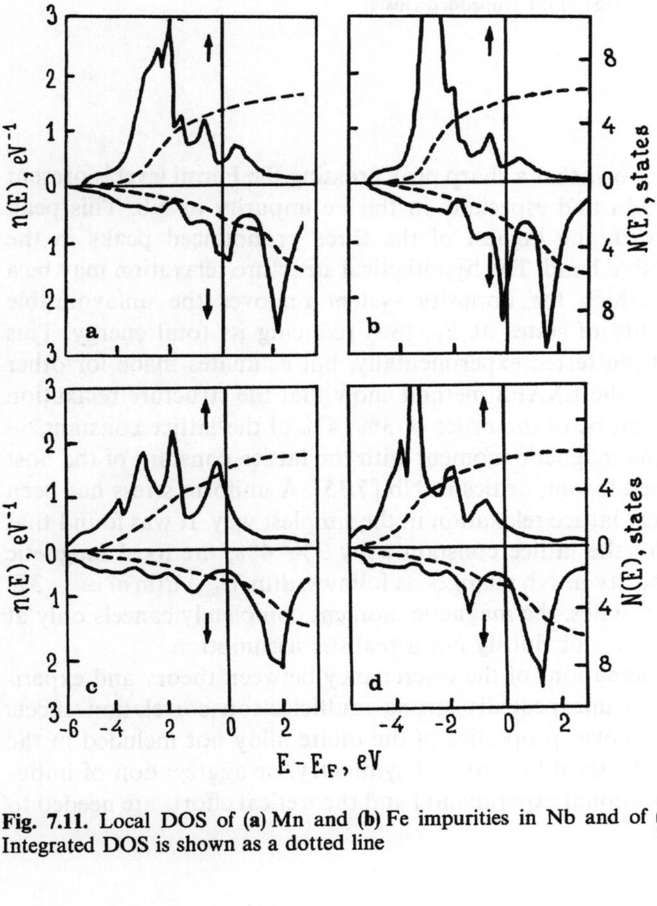

Fig. 7.11. Local DOS of **(a)** Mn and **(b)** Fe impurities in Nb and of **(c)** Fe and **(d)** Mn in Mo. Integrated DOS is shown as a dotted line

Table 7.4. Calculated properties of 3*d* impurities in Nb and Mo

Host		Impurities						
		Ti	V	Cr	Mn	Fe	Co	Ni
Mo	q_{s+p}	1.49	1.69	1.82	1.95	2.08	2.15	2.15
	q_d	2.64	3.66	4.73	5.67	6.63	7.69	8.81
	M_0, μ_B	0	0	0	3.20	3.41	3.42	0
	M_1, μ_B	0	0	0	−0.02	0.05	0.10	0
	M_{cl}, μ_B	0	0	3.03	3.79	3.25	0	
	ΔQ_{cl}	0.00	−0.02	−0.03	−0.02	−0.02	−0.03	−0.03
Nb	q_{s+p}	1.61	1.72	1.91	2.07	2.17	2.19	2.27
	q_d	2.65	3.70	4.75	5.67	6.70	7.81	8.85
	M_0, μ_B	0	0	2.33	3.53	3.20	1.71	0
	M_1, μ_B	0	0	−0.09	−0.12	−0.10	−0.05	0
	M_{cl}, μ_B	0	0	1.64	2.60	2.38	1.32	0
	ΔQ_{cl}	−0.02	−0.04	−0.04	−0.05	−0.05	−0.06	−0.06

q_{s+p} and q_d are the number of electrons in the *s*, *p* and *d* states of the impurity atom;
M_0 is the local magnetic moment at the impurity site;
M_1 is the magnetic moment of the nearest neighbours to the impurity;
M_{cl} is the total magnetic moment of the 15-atom embedded cluster including the impurity;
ΔQ_{cl} is the uncompensated charge of the embedded cluster

impurity. Figure 7.11 shows that a sharp peak crossing the Fermi level is present in the local DOS of Mn and especially in the Fe impurity in Nb. This peak coincides in energy with the highest of the three pronounced peaks in the occupied part of the Nb *d* band. The hypothetical structure relaxation may be a mechanism through which the impurity system removes the unfavourable situation of high density of states at E_F, thus reducing its total energy. This assumption has not been tested experimentally, but estimates made for other similar dilute alloys by the EXAFS method show that the structure relaxation around the impurity can be of the order of 3%–4% of the lattice constant.

The variation of the magnetic moment with the lattice constant of the host crystal has been studied for impurities in Nb [7.35]. A uniform stress has been used to imitate the local lattice relaxation in the simplest way. It was found that for a uniform stress of the lattice constant a by 3%–4%, the local magnetic moment of an Fe impurity in Nb changes as follows: $d(\ln M_{loc})/d(\ln a) \simeq -3.8$. According to these estimates, the magnetic moment completely cancels only at as large a stress as 10%, – obviously not a realistic assumption.

Other possible explanations of the discrepancy between theory and experiment can be related to unexpectedly strong multielectron correlation effects [7.36], or to some unknown properties of the dilute alloy not included in the calculation, for example, local lowering of symmetry, or aggregation of impurities. It seems that additional experimental and theoretical efforts are needed to clear up the situation.

7.2.3 Impurities in Early Transition Metals (Ti, Zr)

Binary alloys of transition metals based on Ti and Zr possess quite interesting magnetic properties, and they give another example of a situation where the numerical electronic structure calculations, similarly very accurate, contradict the experimental evidence.

Magnetic susceptibility measurements on the Ti + 0.2 at % Mn alloy [7.37] estimated $3.5\mu_B$ for the magnetic moment of a Mn impurity. This value was found to be very sensitive to the impurity concentration; the extrapolation of data for a number of alloys to the dilution limit led to the conclusion that an isolated impurity possesses a magnetic configuration with integer spin $S = 2$ [7.38]. This means that there is virtually no hybridization of the impurity d level with the host metal d band.

For Ti–Fe alloys, it is well known that magnetic moments appear only at 70 at % of Fe (for instance, the ordered alloy TiFe is non-magnetic). A special study of the magnetic properties of very dilute TiFe alloys [7.39] also did not find any traces of magnetic moments in Fe atoms.

Table 7.5. Electronic structure parameters for 3*d* impurities in the hcp and bcc phases of Zr as calculated by the LMTO–Green function method

Host crystal structure		Impurities						
		Zr	Nb	Mo	Tc	Ru	Rh	Pd
hcp	S[a.u.]	3.35	3.23	3.15	3.03	2.95	2.88	2.85
	q_{s+p}	1.34	1.28	1.24	1.12	1.02	0.96	1.00
	q_d	2.66	3.72	4.76	5.88	6.98	8.04	9.00
	$n(E_F)$ [Ry^{-1}]	15.2	36.2	47.8	33.6	35.9	15.9	6.9
	ΔE[m Ry]	0	− 15	− 13	15	34	49	58
		Impurities						
		Ti	V	Cr	Mn	Fe	Co	Ni
hcp	S[a.u.]	3.15	3.06	3.01	3.04	2.95	2.77	2.69
	q_{s+p}	1.40	1.33	1.36	1.60	1.32	0.97	0.94
	q_d	2.60	3.67	4.64	5.40	6.48	8.03	9.06
	$n(E_F)$[Ry^{-1}]	23.8	31.1	14.2	11.6	46.1	40.3	10.0
	$M_0[\mu_B]$	0.00	2.35	3.75	3.96	3.07	0.91	0.01
	ΔE[m Ry]	30	45	51	27	35		
bcc	S[a.u.]	3.18	3.10	3.07	3.10	3.02	3.90	2.79
	q_{s+p}	1.38	1.33	1.40	1.65	1.39	0.98	0.95
	q_d	2.62	3.67	4.60	5.35	6.41	8.02	9.05
	$n(E_F)$[Ry^{-1}]	21.2	28.9	19.5	18.1	37.7	16.1	13.0
	$M_0[\mu_B]$	0.00	2.45	3.66	3.86	2.97	0.04	0.00
	ΔE[m Ry]	18	31	30	9	16		

S is the radius of the electrically neutral atomic sphere around the impurity;
q_{s+p} and q_d are the number of electrons in the s, p and d states of the impurity atom;
$n(E_F)$ is the local DOS at the Fermi level;
M_0 is the local magnetic moment at the impurity site;
ΔE is the impurity embedding energy

First-principles calculations for impurities in Ti have been performed by the scattered wave cluster method [7.40] and the LMTO–Green function method [7.41]. In the latter calculation, the matrix elements of the one-centre Green function were calculated for the true hexagonal structure, and then small off-diagonal elements were neglected in subsequent iterations of the impurity potential towards self-consistency.

Calculated parameters of the electronic structure of impurities in the hcp phase and in the high-temperature bcc phase of Zr and Ti are listed in Tables 7.5, 6 [7.41]; and corresponding densities of states are shown in Figs. 7.12, 13. In all cases, no features of resonance type are present in the local DOS; the effects of hybridization between the impurity *d* state and the host metal *d* band play a minor role in determining the spectrum of impurity states. The overall shapes of local DOS for Fe and Mn in Ti are very similar. In the "spin-up" density of states, a pronounced peak is formed near the bottom of the *d* band, while a broad spectrum of "spin-down" states reproduces the main features of the host *d* band. This is due to a peculiar position of spin-split *d* levels of the impurtiy atom with respect to the Ti *d* band. In both Fe and Mn impurities, intra-atomic exchange interaction induces large splitting of *d* levels; the "spin-up" level is situated below the Ti *d* band and only slightly hybridizes with the latter, while the "spin-down" level falls in the middle of the host *d* band and gives a uniform admixture to the different states which belong to it. Relatively small hybridization of the impurity state with the host band reveals considerable localization of the "spin-up" states and hence of the magnetic moment as a whole. Nevertheless, the LMTO–Green function calculation does not support the above-cited experimental estimates for the impurity level width, nor does it provide a criterion for a discrete level to be formed within the host metal *d* band.

Table 7.6 shows that the calculated value of the local magnetic moment of Mn in Ti is lower than the experimentally determined value of $4\mu_B$. An even worse discrepancy occurs with an Fe impurity, for which the magnetic solution with a considerable magnetic moment turned out to be extremely stable and not sensitive to the details of the calculation scheme. This discrepancy reveals the importance of some factor not included in the calculation.

Table 7.6. Electronic structure parameters for 3*d* impurities in the hcp and bcc phases of Ti as calculated by the LMTO–Green function method

Host crystal structure		Impurities					
		V	Cr	Mn	Fe		Co
bcc	S[a.u.]	2.96	2.90	2.88	2.76	2.74	2.70
	$n(E_F)$[Ry^{-1}]	27.1	26.2	25.4	27.2	31.8	46.2
	$M_0[\mu_B]$	0.23	2.41	2.8	2.06	0.0	0.0
hcp	S[a.u.]	2.95	2.90	2.90	2.84	2.81	2.79
	$n(E_F)$[Ry^{-1}]	30.8	18.7	13.6	28.5	3.21	34.1
	$M_0[\mu_B]$	0.23	2.89	3.36	2.74	0.2	0.12

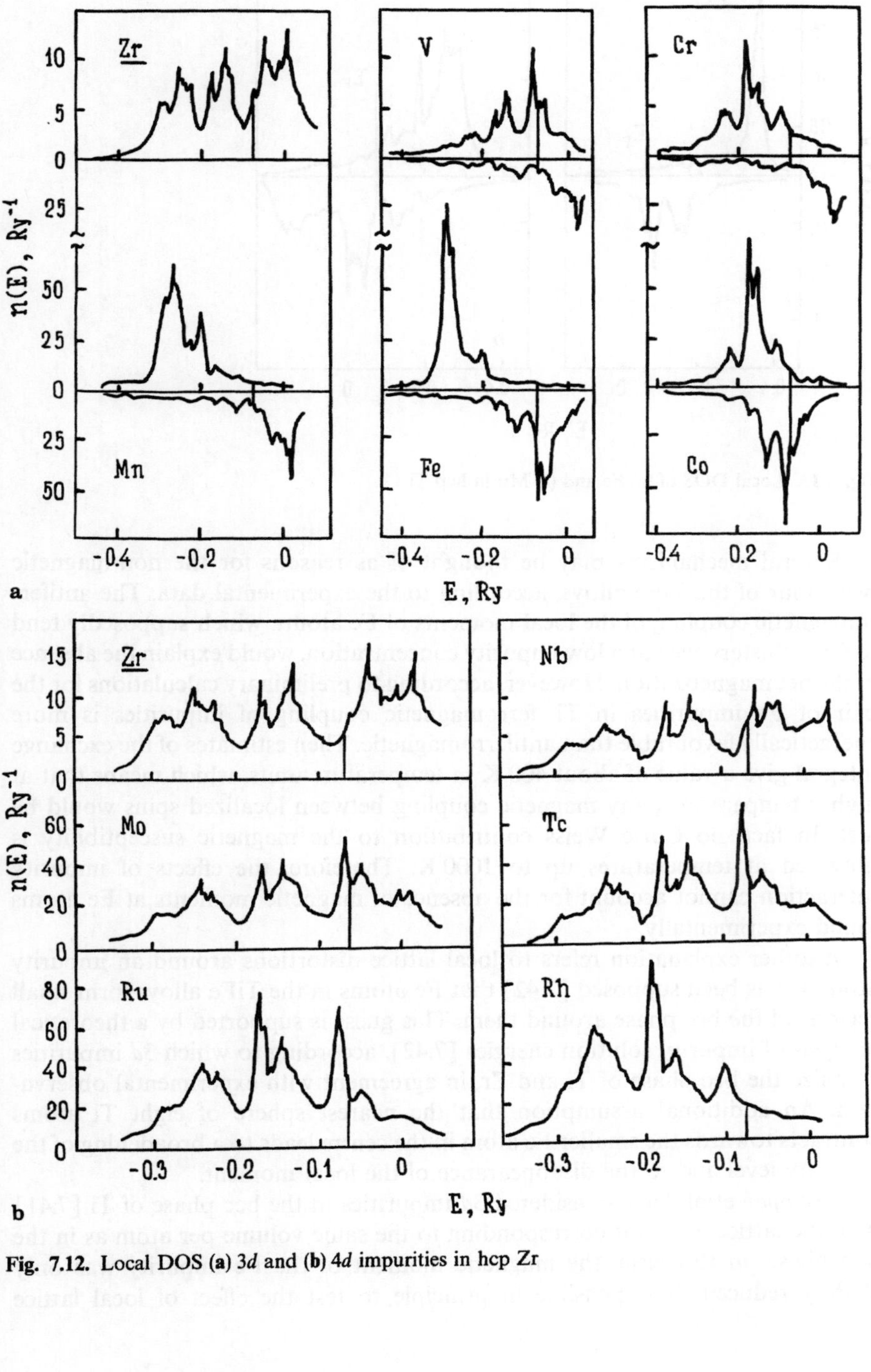

Fig. 7.12. Local DOS (a) 3*d* and (b) 4*d* impurities in hcp Zr

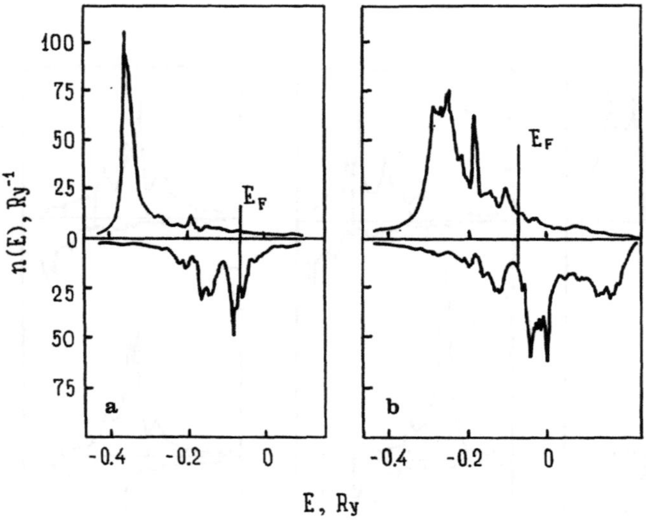

Fig. 7.13. Local DOS of (**a**) Fe and (**b**) Mn in hcp Ti

Several mechanisms may be thought of as reasons for the non-magnetic behaviour of the TiFe alloys, according to the experimental data. The antiferromagnetic coupling of the local moments of Fe atoms, which supposedly tend to form clusters even at a low impurity concentration, would explain the absence of the net magnetization. However, according to preliminary calculations for the pair of Fe impurities in Ti, ferromagnetic coupling of impurities is more energetically favourable than antiferromagnetic. Then estimates of the exchange integral give a value of about 400 K in temperature units, which means that at higher temperatures any magnetic coupling between localized spins would be lost. In fact, no Curie–Weiss contribution to the magnetic susceptibility is observed at temperatures up to 1000 K. Therefore, the effects of impurity interaction cannot account for the absence of magnetic moments at Fe atoms found experimentally.

Another explanation refers to local lattice distortions around an impurity atom. It has been supposed [7.42] that Fe atoms in the TiFe alloys form small regions of the bcc phase around them. This guess is supported by a theoretical analysis of impurity solution energies [7.42], according to which 3*d* impurities stabilize the bcc phase of Ti and Zr, in agreement with experimental observation. An additional assumption that the nearest sphere of eight Ti atoms contracts towards the smaller Fe atom in the centre leads to a broadening of the impurity level and to the disappearance of the local moment.

Antropov et al. have considered 3*d* impurities in the bcc phase of Ti [7.41] with the lattice constant corresponding to the same volume per atom as in the hcp phase; in this case, the magnetic moment of the Fe impurity was only slightly reduced. It is possible in principle to test the effect of local lattice contraction in the KKR–Green function calculations, as has been done for impurities in Cu [7.43]. However, no such calculations have so far been performed for Ti-based systems.

7.3 Impurities in Magnetic Metals

From the computational point of view, calculation of the electronic structure of impurities in magnetic hosts is no more complicated than the calculations discussed above. However, as the presence of spontaneous magnetization (in ferromagnets) introduces new important interactions, a variety of quite specific problems – like the variation of the hyperfine field due to the presence of impurities, or the prediction of correct orientation of the impurity magnetic moment with respect to that of the host–must be solved. Many papers deal with ab initio calculations for impurities in ferromagnets. The first calculation known to us is one for the impurity states in Fe within the LMTO–Green function formalism [7.44]. These single-site calculations were not really self-consistent because the potential in the impurity atomic sphere was shifted rigidly until the Friedel sum rule was satisfied. Later on, as more accurate single-site calculations taking into account realistic adjustment of the impurity potential appeared ([7.45], for impurities in Ni), it became clear that the potential perturbation due to the presence of an impurity involves several spheres of neighbours. In more recent papers, the Friedel sum rule analysis was used to estimate the spatial extent of the potential perturbation. Calculations for $3d$ and $4d$ impurities in Ni have been performed with an embedded cluster of 55 atoms [7.46].

We discuss below first the calculated data available for impurities in iron, the subject in especially many papers. Then we review results for Ni- and Co-based systems, and discuss the impurities in antiferromagnets and intermetallic alloys.

7.3.1 Fe-Based Impurity Systems

As is well known, the ground state of ferromagnetic iron has a bcc structure. The high-temperature fcc phase of Fe can be stabilized at lower temperatures by alloying; it is the basic structure for many iron-containing alloys, for example, the Invar systems. Therefore, the problem concerning the effect of alloying on the stability of different magnetic structures is of practical interest, and in this respect the electronic structure calculations for impurities in both fcc and bcc Fe can be very helpful.

A systematic study of the electronic structure of s, p impurities in bcc Fe has been carried out [7.47] on the basis of the KKR–Green function calculation. Figure 7.14 shows the local densities for impurities along with the energy-resolved spin densities at the nucleus $\rho_c(E) = -1/\pi \, \mathrm{Im}\{G_\sigma(0, 0, E)\}$. The latter property is closely related to the hyperfine field at the nucleus and hence to the isomer shift.

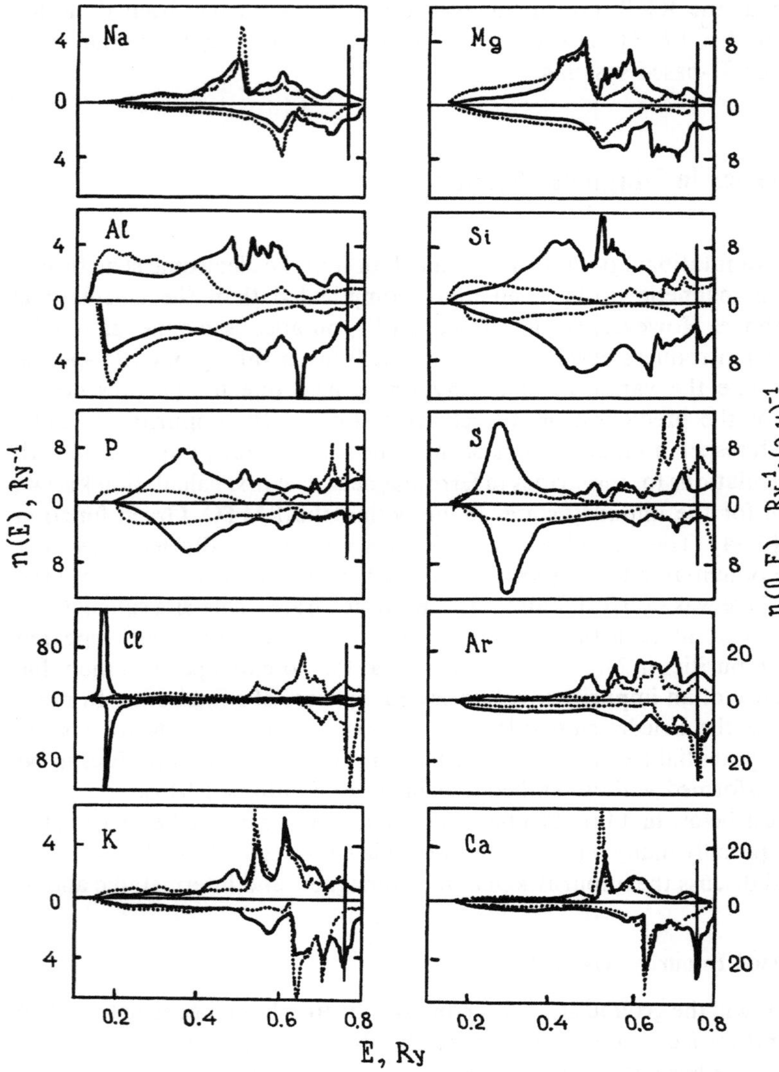

Fig. 7.14. Local DOS of *s*, *p* impurities in bcc Fe (*solid lines*) and spatial spin density at impurity nuclei (*dotted lines*)

Corresponding results for 3*d* and 4*d* impurities in Fe have been obtained [7.48]. Figure 7.15 shows the local DOS of 3*d* impurities as calculated by the LMTO–Green function method [7.49].

As Figs. 7.14, 15 show, strong hybridization of impurity states with those of the ferromagnetic host is a peculiar feature of this system. As a result, the impurity DOS exhibits non-regular behaviour in the row of impurities, with

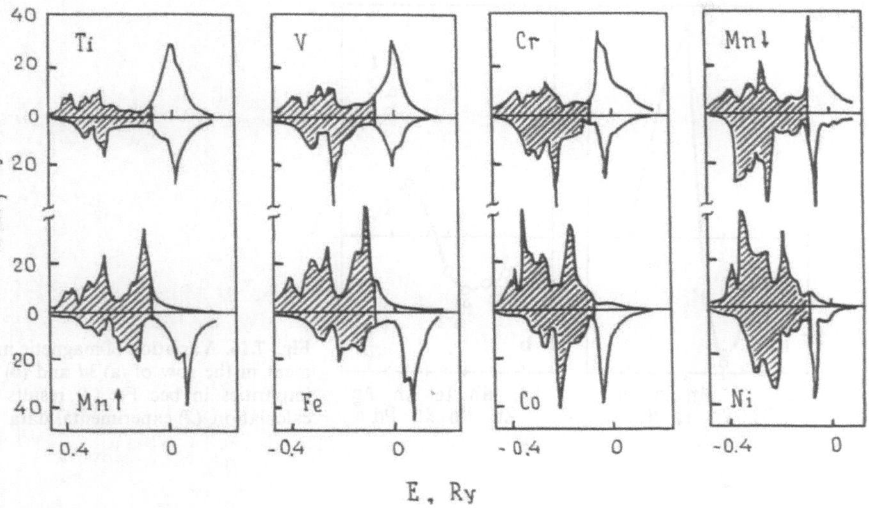

Fig. 7.15. Local DOS of 3*d* impurities in bcc Fe

noticeable induced spin polarization in all cases. For impurities from the second half of the 3*d* (or 4*d*) series, no features resembling resonance peaks are seen in the density of states. Instead, the impurity DOS mostly reproduces that of the host, with the relative intensities, of peaks changed. The ferromagnetic splitting of bands in the impurity DOS is retained, so that the magnetic moment at the impurity site is aligned parallel to the host magnetization. A quite different shape of the local DOS is typical for the impurities of the first half of the row (Ti, V), for which a resonance peak appearing 0.1–0.3 Ry above the Fermi level is the dominating feature. This resonance peak, which is present in both spin-up and spin-down densities of states, originates from high-lying *d* levels of impurity atoms. These impurity levels are closer in energy to the spin-down host states, which are moved upwards by the ferromagnetic splitting, and more effective hybridization with them leads to the predominant occupation of spin-down states at the impurity site. Therefore, an inversion of the local moment with respect to that of the host occurs for impurities of the first half of the 3*d* series. As is shown below, this situation is typical for impurities in all ferromagnetic hosts.

Figure 7.16 shows calculated values of local magnetic moments for 3*d* and 4*d* impurities. Reasonably good agreement should be noted in almost all cases. However, noticeable discrepancies in both the magnitude and the sign of the magnetic moment are reported for the critical region near the middle of the 3*d* (4*d*) row, where the magnetic moment abruptly changes its direction. We refer to the following estimates of the calculated magnetic moment of Mn in Fe: $1.6\mu_B$ [7.50], $0.6\mu_B$ [7.48], and $-2.3\mu_B$ and $1.6\mu_B$ [7.44]. Apparently, the results obtained for this critical region are extremely sensitive to the details of the calculation scheme applied. *Akai* et al. [7.48, 50] used the KKR–Green function

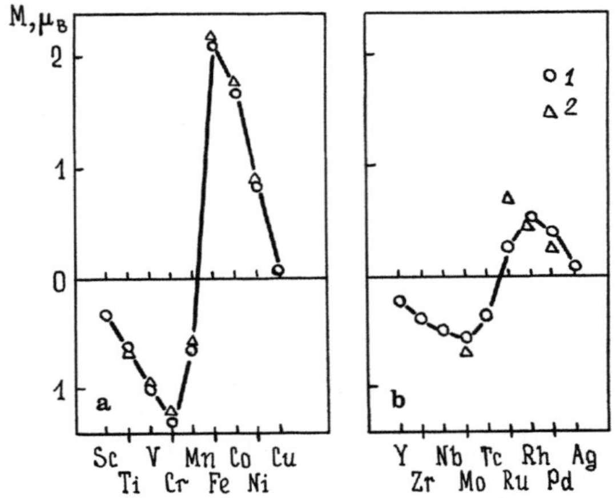

Fig. 7.16. Variation of magnetic moment in the row of (**a**) 3*d* and (**b**) 4*d* impurities in bcc Fe: (*1*) results of calculation, (*2*) experimental data

method with the "embedded cluster" in the latter case, whereas *Anisimov* et al. [7.49] applied the LMTO–ASA–Green function method. There were also differences in the exchange-correlation potential chosen: one study [7.50] used the prescriptions by von Barth–Hedin; the other [7.49] used those by Gunnarsson–Lundqvist (Chap. 4). This latter difference could be crucial for determining of the values of magnetic moments since, as was pointed out [7.45], the calculated magnetic polarization is extremely sensitive to the detailed shape of the exchange-correlation potential.

According to *Anisimov* et al. [7.49], two self-consistent solutions, corresponding to different orientations of the impurity moment, have been found for Mn in Fe. Such a possibility was first mentioned for the case of Mn in Ni [7.45]. An analysis of the preferential stability of the two solutions was not performed [7.45], and one of these solutions was omitted as contradicting experimental evidence. *Anisimov* et al. [7.49] calculated the effective exchange parameter (5.66) for both self-consistent solutions to test their stability against deviations of local moments from collinearity. Table 7.7 lists the values of J_0 and the magnetic moments obtained [7.49].

For both spin configurations found for Mn in bcc Fe, the exchange parameters are positive. This means that none of these solutions is a false one but, on the contrary, they correspond to two local minima of the total energy. This observation favours the assumption that dynamic fluctuations are possible between parallel and antiparallel spin configurations of the Mn impurity. This would explain a broad range of experimental estimates of the magnetic moment [7.51] for this system.

It makes sense to plot the exchange parameter J_0 as a function of the upper integration limit in the (5.66) to get some idea of how this value depends on the band filling and how sensitive it is to small variations of the electronic structure. These plots [7.49] are given in Fig. 7.17. The vertical line gives the actual

Table 7.7. Calculated values of local magnetic moments M [μ_B] and effective exchange parameters J_0 [mRy] for 3*d* impurities in ferromagnetic metals

Host		Ti	V	Cr	Mn↓	Mn↑	Fe↓	Fe↑	Co	Ni
bcc Fe	M	−0.82	−1.13	−1.60	−2.30	1.60	—	2.25	1.49	0.73
	J_0	1.2	2.9	12.4	9.3	1.5	—	17.1	15.4	7.9
fcc Fe	M	−0.90	−1.36	−2.26	−2.96	2.48	−2.72	2.43	1.22	0.56
	J_0	1.3	10.9	16.0	12.1	−7.2	−0.7	11.2	8.9	4.6
hcp Co	M	−0.52	−0.90	−1.73	−2.73	2.03	−2.40	2.48	1.51	0.62
	J_0	2.4	5.4	10.2	10.1	5.3	−17.0	15.3	13.5	5.8
fcc Co	M	−0.50	−1.16	−1.78	−2.88	2.66	−2.31	2.50	1.49	0.62
	J_0	2.1	8.1	10.6	−27	12.6	−28.4	15.3	14.7	5.8
fcc Ni	M	−0.11	−0.42	−0.79	−2.87	2.89	−3.10	2.72	1.51	0.52
	J_0	0.2	1.2	1.8	−4.6	8.8	−18.0	11.6	9.3	2.9

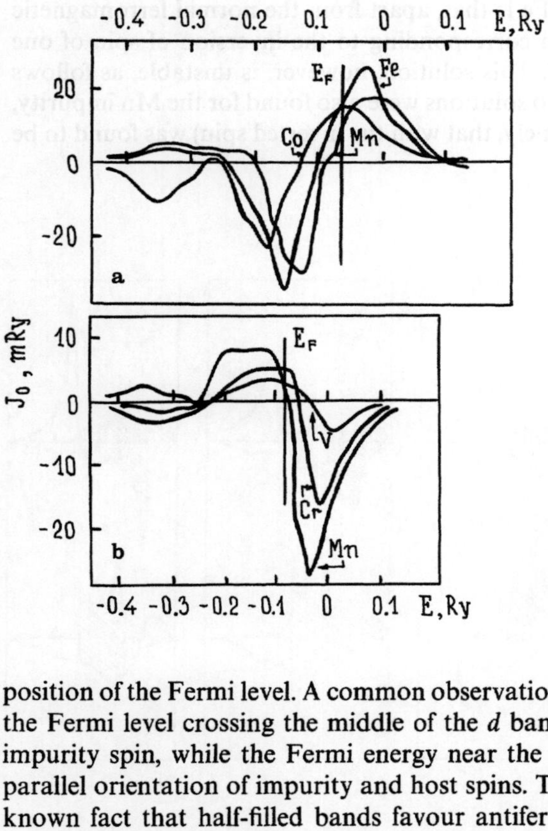

Fig. 7.17. Effective exchange parameter J_0 as a function of E_F for 3*d* impurities in bcc Fe: (**a**) for impurities with "positive" magnetic moments, (**b**) for impurities with "negative" magnetic moments

position of the Fermi level. A common observation concerning these plots is that the Fermi level crossing the middle of the *d* band favours the case of inverted impurity spin, while the Fermi energy near the top of the band stabilizes the parallel orientation of impurity and host spins. This is consistent with the well-known fact that half-filled bands favour antiferromanetic ordering (Chap. 2), while almost filled or almost empty bands tend to ferromagnetism because the double exchange dominates.

Note that the lowest part of the J_0 curve, seemingly favouring ferromagnetic ordering, in practice never becomes real because the band filling corresponding to it is too low to give rise to any magnetic moment at all.

Of all the curves shown in Fig. 7.17, the one corresponding to Mn seems to be the most sensitive to the peculiarities of the calculation scheme, since it almost crosses the zero point at the Fermi level. This means that the mgnetic behaviour of Mn impurity in Fe should be revised in more accurate calculations.

As was mentioned in Chap. 5, the fcc structure of pure Fe occurs only at temperatures above the Curie point. However, at lower temperatures this structure can be stabilized by alloying. Therefore, a theoretical study of magnetic impurities in fcc Fe can be of practical interest.

The electronic structure of 3*d* impurities in ferromagnetic fcc iron has been calculated by the LMTO-Green function method [7.49]. The lattice constant chosen correspond to the "high-spin" state of ferromagnetic iron [7.52]. Figure 7.18 shows the local DOS of impurities obtained [7.49]. Table 7.7 gives the values of magnetic moments and exchange parameters. An important difference from the case of bcc Fe is that, apart from the normal ferromagnetic solution, an additional solution corresponding to the inversion of spin of one atom was found for "Fe in Fe". This solution, however, is unstable, as follows from the negative sign of J_0. Two solutions were also found for the Mn impurity, and again only one of them (namely, that with the inverted spin) was found to be a true stable solution.

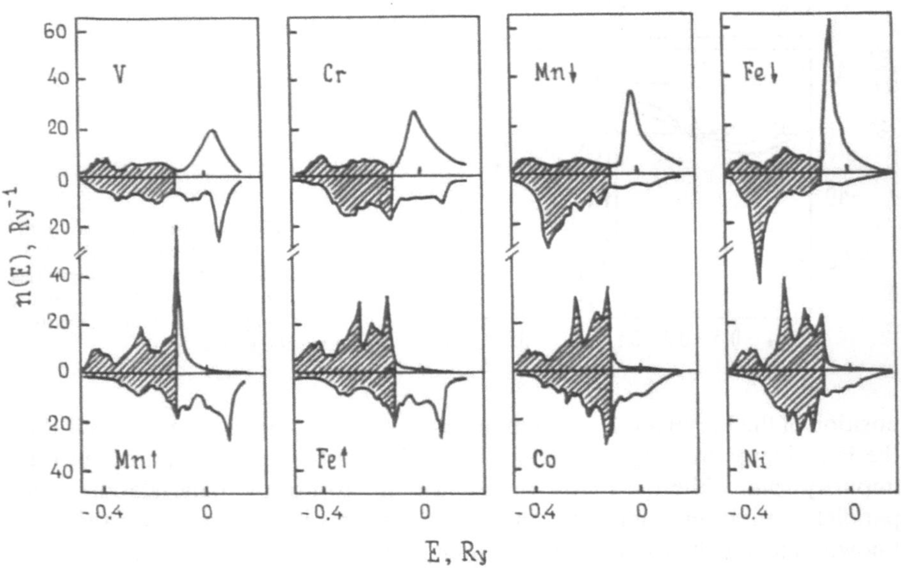

Fig. 7.18. Local DOS of 3*d* impurities in fcc Fe

Comparison of the exchange parameter values J_0 (Table 7.7) makes it possible to conclude that the Curie temperature is considerably lower for the fcc phase than for bcc, despite the opposite relation between the values of magnetic moments. This is consistent with experimental measurements of magnetic properties of γ-Fe precipitates in copper [7.53].

7.3.2 Ni-Based Systems

Starting from the papers [7.54, 55] devoted to magnetization measurements in dilute Ni alloys, a number of both experimental and theoretical studies have been undertaken to understand better the electronic structure of these materials [7.56]. The most reliable first-principles calculations concerning *s*, *p* and 3*d*, 4*d* impurities in nickel were performed [7.46] within the KKR–Green function formalism. The calculation was performed for an embedded cluster of 55 atoms including the impurity; i.e. the perturbation of potential and magnetic polarization within four spheres of neighbours was considered. The single-site LMTO–Green function calculation for 3*d* impurities in nickel has been performed along with similar calculations for other ferromagnets [7.49]. Table 7.7 lists the magnitude of magnetic moments and the values of effective exchange parameters obtained in these calculations.

All papers mentioned neglected possible shifts of nickel-atom relaxation around impurities which are of different sizes. Such structure distortions can be as large as 1.4% of the mean interatomic distance, giving rise to a change in the magnetic-moment value of $\sim 0.1\mu_B$ [7.57].

Figure 7.19 shows local DOS of 3*d* impurities in Ni according to the calculation [7.49]. For the impurities of the beginning of the 3*d* row (V, Cr), pronounced resonances above the Fermi level are seen, and an inversion of the impurity moment takes place, as was discussed above for impurities in Fe. For subsequent impurities, the virtual *d* level for spin-up and then for spin-down states gets lower and falls into the continuum of the host metal *d* band. For the later impurities (Co to Cu), the local density of states resembles that of pure Ni, and the impurity magnetic moments are aligned parallel to the host magnetization.

As Table 7.7 shows, pairs of self-consistent solutions corresponding to parallel and antiparallel orientations of the magnetic moment were found for Mn and Fe impurities. This is consistent with the results of the earlier KKR–Green function calculation [7.58] where single-site impurity calculations were performed so that the nuclear charge of the impurity atom was allowed to be non-integer. The coexistence of parallel and antiparallel solutions was found for $24.2 < Z < 26.5$, i.e. for the region embracing Z values for Mn and Fe (Fig. 7.20). The comparative stability of the solutions was not discussed [7.58].

This analysis has been done based on the values of effective exchange parameter J_0 [7.49] (Table 7.7), from which antiparallel solutions for both Mn and Fe impurities appeared to be unstable. Figure 7.21 shows the J_0 value dependence on the Fermi level position [7.49], analogous to Fig. 7.17.

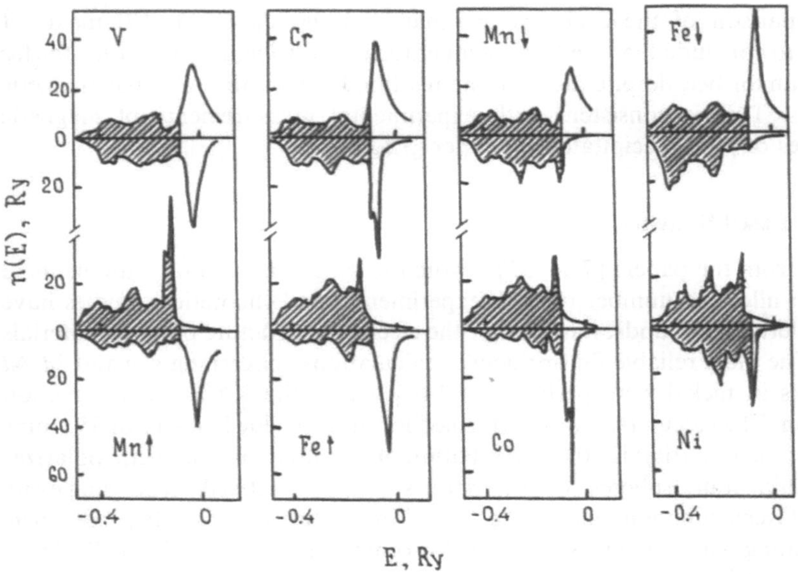

Fig. 7.19. Local DOS of 3*d* impurities in fcc Ni

The experimental data on the DOS at the impurity site, such as X-ray emission or X-ray photoelectron spectra, are not practically available for 3*d* impurities in Ni. However, the X-ray emission spectra for dilute NiMn alloy have been obtained [7.59], discussed, and compared with a cluster calculation [7.60], but the experimental energy resolution in the X-ray spectra was not sufficient to obtain any details about DOS fine structure.

Experimental values of the local magnetic moments of impurities shown in Fig. 7.20 are determined mostly by unpolarized or polarized neutron scattering. The results of the calculation agree well with the experimental data, with the most noticeable discrepancies for chromium. The only indication of parallel orientation of the Cr magnetic moment has been reported [7.61] with rather broad error bars: $0.2 \pm 0.6\mu_B$. Therefore this does not actually contradict the calculation results. Nevertheless, the calculation-based estimates of the Cr impurity magnetic moment have proved sensitive to the computational details and need therefore to be specified in further calculations.

The magnetic moments induced on the 4*d* impurities embedded in ferromagnetic Ni have also been calculated [7.46]. The change of these magnetic moments in the 4*d* row obeys the same trends as for 3*d* impurities; i.e. the local moment orientation is antiparallel at the beginning of the row and changes to parallel in the middle. No double solutions were found for 4*d* impurities. The magnitudes of the magnetic moments are systematically lower than for 3*d* impurities.

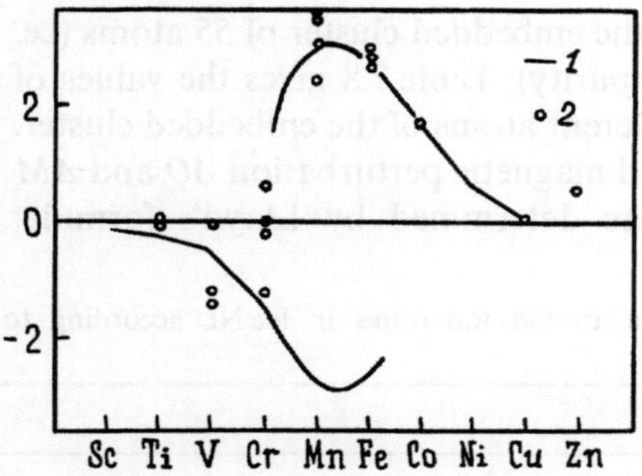

Fig. 7.20. Local magnetic moments of 3*d* impurities in Ni: (*1*) calculation results, including the case of non-integer nuclear charge values, (*2*) experimental data

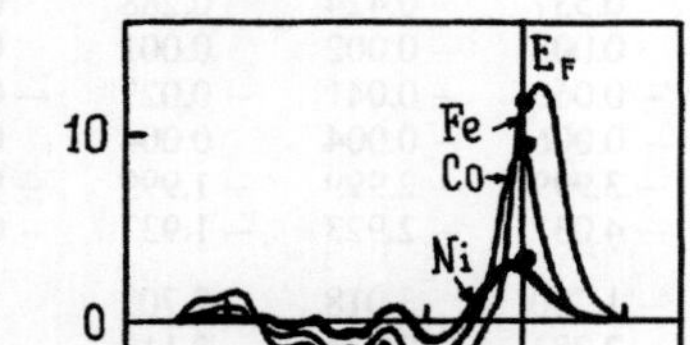

Fig. 7.21. Effective exchange parameter J_0 as a function of E_F for 3*d* impurities in fcc Ni: **(a)** for impurities with "positive" magnetic moments, **(b)** for impurities with "negative" magnetic moments

Experimental data on the magnetic-moment values are available only for several of the 4*d* impurities. For Rh in Ni, there is a striking disagreement between calculational result and the experimental estimate of $m_{Rh} = 2\mu_B$ [7.62]. This experimental value seems rather doubtful because it exceeds the magnetic moment found for Co. It is well understood, and consistent with other experimental results, that magnetic moments of 4*d* impurities are smaller than the moments of their 3*d* analogues.

The spatial distribution of charge and magnetic density around impurities in Ni has been studied [7.46] with the KKR–Green function method as the potentials were allowed to change within the embedded cluster of 55 atoms (i.e. four spheres of neighbours around the impurity). Table 7.8 gives the values of charge and magnetic perturbations for different atoms of the embedded cluster. The asymptotic values of global charge and magnetic perturbation ΔQ and ΔM due to the inclusion of the impurity are determined by Lloyd's formula.

Table 7.8. Charge and magnetic-moment perturbations around impurities in fcc Ni, according to the KKR–Green function calculation [7.46]

	Impurities								
	K	Ca	Sc	Ti	V	Cr	Mn	Fe	Co
ΔQ_0	−8.888	−8.599	−7.877	−6.817	−5.666	−4.478	−3.376	−2.265	−1.119
ΔQ_1	0.000	0.752	1.027	0.932	0.750	0.537	0.424	0.288	0.124
ΔQ_2	0.106	0.071	0.047	0.026	0.012	0.008	−0.002	0.001	0.001
ΔQ_3	−0.148	−0.164	−0.146	−0.109	−0.081	−0.060	−0.041	−0.027	−0.013
ΔQ_4	−0.080	−0.065	−0.054	−0.032	−0.013	−0.006	−0.004	0.004	0.009
ΔQ_{cl}	−9.010	−8.005	−7.003	−6.000	−4.998	−3.999	−2.999	−1.999	−0.999
ΔQ	−9.898	−8.135	−7.069	−6.035	−5.008	−4.031	−2.923	−1.937	−0.957
M_0	−0.014	−0.067	−0.133	−0.272	−0.562	−1.700	3.018	2.703	1.702
ΔM_0	−5.598	−0.650	−0.717	−0.856	−1.145	−2.283	2.434	2.119	1.119
ΔM_1	−0.034	−0.124	−0.164	−0.193	−0.204	−0.180	−0.033	−0.011	0.004
ΔM_2	−0.011	−0.035	−0.042	−0.047	−0.046	−0.048	0.024	0.003	−0.003
ΔM_3	−0.004	−0.007	−0.012	−0.017	−0.020	−0.018	0.002	0.001	0.002
ΔM_4	−0.007	−0.000	−0.006	−0.012	−0.017	−0.012	−0.011	−0.014	−0.010
ΔM_{cl}	−1.256	−2.528	−3.299	−4.013	−4.546	−5.300	2.109	1.862	1.067
ΔM	−1.169	−2.484	−3.257	−3.988	−4.557	−5.319	2.140	1.813	1.017

	Impurities						
	Cu	Zn	Ga	Ge	As	Se	Br
ΔQ_0	1.026	1.858	2.684	3.637	4.693	5.810	6.934
ΔQ_1	−0.016	0.187	0.414	0.485	0.452	0.361	0.241
ΔQ_2	0.003	0.007	0.008	−0.001	−0.000	0.016	0.033
ΔQ_3	−0.001	−0.030	−0.051	−0.094	−0.117	−0.117	−0.137
ΔQ_4	−0.012	−0.022	−0.064	−0.029	−0.028	−0.084	−0.085
ΔQ_{cl}	1.000	2.000	2.991	3.999	5.000	5.986	6.986
ΔQ	0.999	2.025	3.166	3.972	4.831	5.867	6.567
M_0	−0.013	−0.052	−0.052	−0.040	−0.014	0.029	0.075
ΔM_0	−0.596	−0.635	−0.636	−0.623	−0.598	−0.554	−0.508
ΔM_1	−0.033	−0.080	−0.126	−0.157	−0.163	−0.143	−0.099
ΔM_2	−0.005	−0.016	−0.024	−0.028	−0.031	−0.033	−0.028
ΔM_3	−0.004	−0.006	−0.011	−0.008	−0.006	−0.007	0.000
ΔM_4	0.000	−0.006	−0.024	−0.014	−0.014	−0.032	−0.030
ΔM_{cl}	−1.119	−1.915	−2.849	−3.027	−3.051	−3.019	−2.238
ΔM	−1.068	−1.871	−2.674	−2.935	−2.956	−2.791	−2.046

ΔQ_i is the change of the total number of valence electrons relative to pure Ni in the impurity MT sphere ($i = 0$) and for atoms of the ith sphere of neighbours to the impurity ($i \neq 0$);
ΔQ_{cl} is the sum of ΔQ_i over the 55-atom embedded cluster;
M_0 is the local magnetic moment at the impurity site;
ΔM_i is the perturbation of the magnetic moment from that of pure Ni for ith neighbours to the impurity atom;
ΔM_{cl} is the total perturbation of the magnetic moment within the 55-atom embedded cluster;
ΔM is the perturbation of the total magnetic moment of the crystal due to the presence of an impurity as estimated by the Friedel sum rule

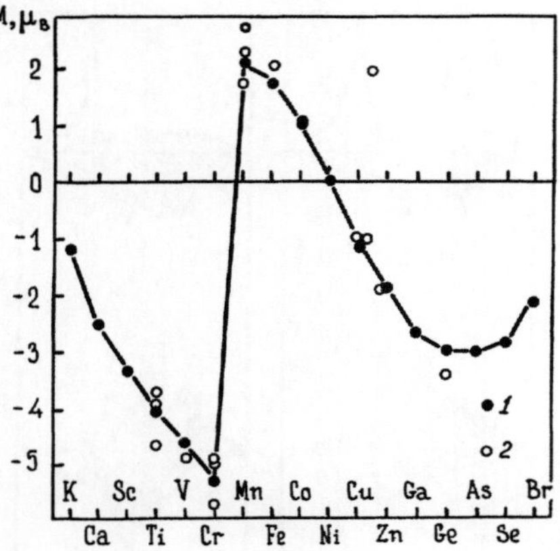

Fig. 7.22. Change of the Ni host magnetization due to alloying with 3*d* impurities: (*1*) calculation results, (*2*) experimental data

Table 7.8 shows that for almost all the impurities considered the charge and magnetic perturbations are localized within four spheres of neighbours. Noticeable exclusions are K and Br impurities, as the perturbation introduced by them is obviously more distant.

The derivative of the magnetization $(dM/dc)|_{c=0}$ was calculated for Ni-based dilute alloys [7.46]. Figure 7.22 shows the variation of this property in the row of elements from K to Br, in comparison with the experimental data. Fairly good agreement supports the prospects of applying LSDF-based methods for studying magnetic excitations.

7.3.3 Co-Based Impurities

The theoretical study of the electronic structure of impurities in cobalt is complicated by the fact that the ground state of Co possesses a hexagonal structure that leads to non-diagonality of the single-site impurity T matrix and hence to larger computational efforts. This is by no means a fundamental problem; however, impurity calculations treating the crystal structure of cobalt without any simplifications have not been reported so far, according to our knowledge.

The Green function of pure Co has been calculated for the correct crystal structure [7.49], but with only the dominating diagonal elements being retained in a one-site scattering path operator matrix for further impurity calculations. Figure 7.23 shows the calculated local densities of states obtained [7.49]; the values of local magnetic moments and the effective exchange parameters are listed in Table 7.7.

Calculations have also been performed for impurities in the fcc phase of Co which is known to exist at high temperatures [7.49].

For both hcp and fcc structures of the host, double solutions corresponding to different orientations of impurity local moments were found for Mn and Fe,

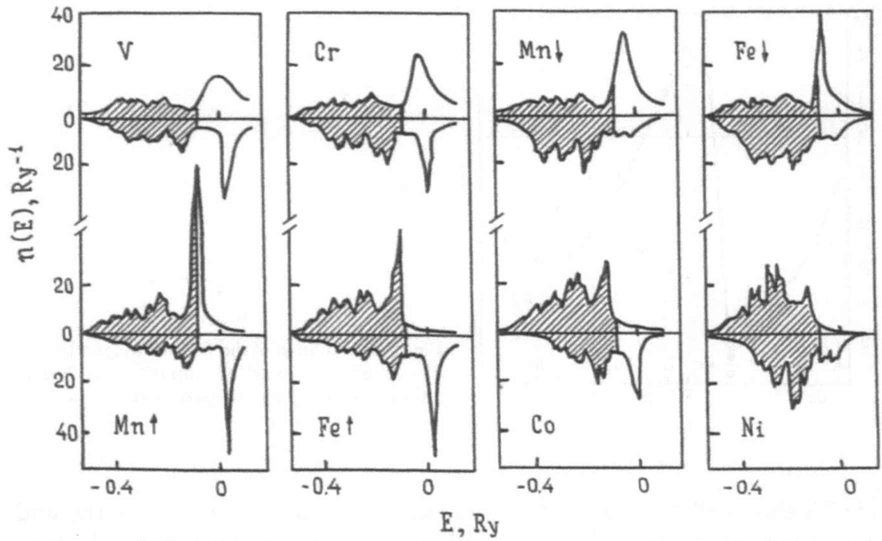

Fig. 7.23. Local DOS of 3*d* impurities in hcp Co

i.e. for the elements of the middle of the 3*d* row where the inversion of the spin moment takes place. For an Mn impurity in hcp Co, both solutions should be retained, as J_0 is positive for them. The value of J_0 is larger for antiparallel orientation of the magnetic moment of an Mn impurity in hcp cobalt. We can assume here that this antiparallel solution corresponds to a sharper minimum of the total energy (because J_0 is proportional to its second derivative over the deviation angles (5.61)). Probably a more stable magnetic configuration than the parallel solution exists.

In the three other systems which exhibit double solutions, only cases of parallel orientation of the impurity moment should be retained as the stable ones.

These conclusions are consistent with the experimental evidence concerning the magnetic state of Mn in Co. As has been mentioned [7.63], on the basis of the NMR data, the Co–Mn coupling in fcc Co is ferromagnetic. The polarized neutron scattering data for hcp CoMn alloy show that the local magnetic moment of Mn in this system is directed antiparallel to the host magnetization, and its magnitude is only $\sim 0.3\mu_B$ [7.64]. It has been supposed [7.65] that such a low value as compared with the NiMn system is due to dynamic fluctuations between parallel and antiparallel orientations of the Mn magnetic moment.

7.3.4 Impurities in Antiferromagnets

Antiferromagnetism of chromium- and manganese-based dilute alloys presents one of the most complex problems of itinerant magnetism theory. As is well known, the antiferromagnetic structure of pure Cr is incommensurate with the crystal lattice and has the form of a spin-density wave with the antiferromagnetism vector $\boldsymbol{Q} = 2\pi/a\ [(1-\delta)00]$, $\delta \simeq 0.05$ [7.66]. This is due to the specific topology of the Fermi surface [7.67]. Small amounts of a 3*d* element (2.5% of Fe or 1.1% of Mn [7.68]), however, make the spin-density wave commensurate, resulting in a simple antiferromagnetic ordering of the B2 type in dilute alloys.

The antiferromagnetic ground state of manganese is also rather complicated, with 58 atoms in four non-equivalent positions in the unit cell [7.69]. This makes experimental determination of the magnetic properties of Mn-based alloys troublesome.

Experimental data available for impurities in Cr are also not numerous. Measurements of the magnetic susceptibility of the CrFe alloy reveal that the iron impurity below the Néel point possesses a magnetic moment of $2\mu_B$, which is slightly reduced at higher temperatures [7.68]. *Ishikawa* et al. stated that the Fe magnetic moment is weakly coupled to the magnetic momens of the host [7.68]. A study of magnetic and transport properties of CrFe and CrCo alloys [7.70] supported their result concerning the weakness of the Fe–Cr magnetic interaction. In contrast to this, strong magnetic interaction was reported for a Co impurity in Cr.

Band calculations of the electronic structure of pure Cr [7.71, 72] show that the (commensurate) antiferromagnetic state has lower energy than the non-magnetic one. The tendency towards antiferromagnetism which is due to the half-filling of the *d* band is reproduced even in small-cluster calculations [7.73]. The energy preference of the non-commensurate structure was not tested in numerical calculations; it is most probably governed by the competition of exchange interactions between nearest and more distant neighbours [7.74].

For Mn, as follows from the band-structure calculation [7.75], the B2 antiferromagnetic phase is lower in energy than both non-magnetic and ferromagnetic phases. More realistic calculations for the true ground state of manganese recently proved feasible [7.76]; the task of calculating the Green function and solving the impurity problem for such a complicated structure is still much too complicated. Meanwhile, the impurity calculations performed for a simple B2 antiferromagnetic phase (the crystal structure of which is identical to that of the high-temperature δ-Mn) may provide some preliminary data.

Self-consistent calculations of the electronic structure and magnetic interactions for 3*d* impurities in antiferromagnetic Cr and Mn have been performed [7.77] by the LMTO–Green function method. Figures 7.24, 25 show the local densities of states; Table 7.9 lists calculated values of magnetic moments and exchange integrals. Starting from Fe (and for Cr in Mn as well), sharp narrow peaks appear in the impurity densities of states near the Fermi level in Cr and Mn. These peaks correspond to the non-bonding impurity states formed in an "antiferromagnetic gap" in the middle of the host conduction band.

Let us now discuss the effective exchange parameter. It is negative for the spin-down configuration of the Fe impurity in Cr; therefore, the "spin-up" solution should be retained as the true one. The exchange bonding between the

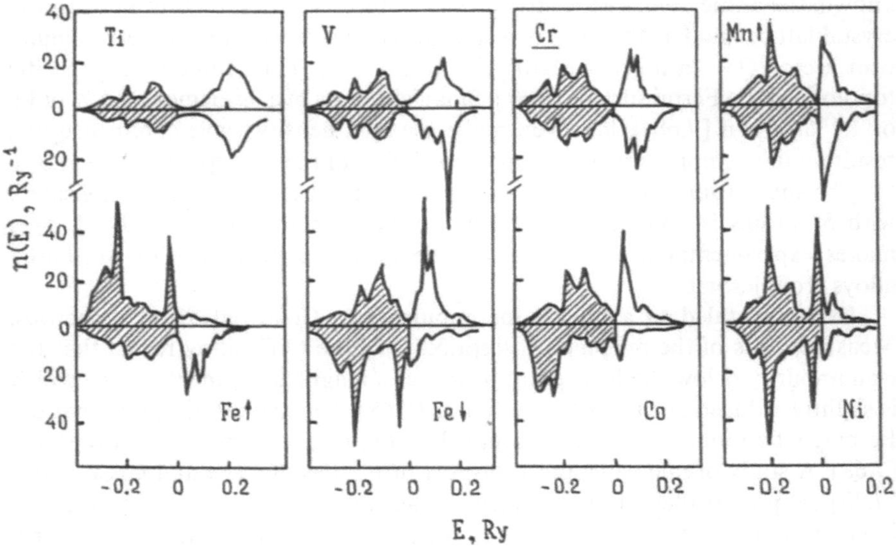

Fig. 7.24. Local DOS of 3*d* impurities in Cr

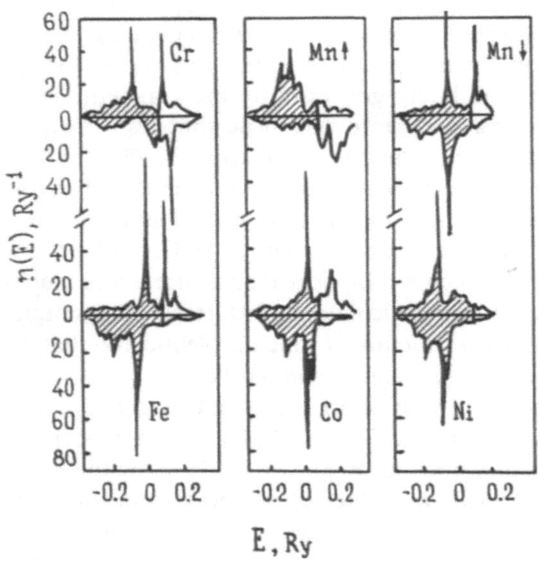

Fig. 7.25. Local DOS of 3*d* impurities in Mn

Fe impurity and the Cr host is, however, rather weak, even compared with that of pure chromium. For the Co impurity, the value of the exchange parameter is considerably greater and agrees with the experimental evidence of stronger exchange coupling in CrFe and CrCo alloys.

Table 7.9. Calculated values of local magnetic moments $M\,[\mu_B]$ and effective exchange parameters J_0 [m Ry] for $3d$ impurities in antiferromagnetic metals

Host		Impurities							
		V	Cr	Mn ↓	Mn ↑	Fe ↓	Fe ↑	Co	Ni
Cr	M	0.36	0.59	0.48	—	2.17	− 1.83	− 1.63	− 0.13
	J_0	1.0	2.5	1.4	—	0.7	− 2.8	6.2	0.4
Mn	M	1.10	2.18	2.98	1.12	—	− 1.86	− 1.34	− 1.37
	J_0	5.6	11.8	− 8.8	− 4.1	—	2.4	8.5	1.8

The fact that both solutions for a Mn "impurity" in Mn (i.e., the cases of the pure antiferromagnetic metal and that with a reversed spin at one site) correspond to negative values of J_0 reveals that neither configuration occurs in reality. The formal existence of two self-consistent solutions means, however, that a magnetic moment at the Mn site is localized rather than itinerant.

7.3.5 Impurities in FeCo Alloys

The electronic structure of impurities in the ferromagnetic ordered alloy FeCo has been studied [7.78]. The FeCo system (at about 70 at % Fe) is known to possess the maximum bulk magnetization among the $3d$ metal-based compounds; this is related to the highest value of the magnetic moment at the Fe atom (about $3\mu_B$) ever detected in iron alloys, which is formed in this system at about 50 at % Fe concentration.

The origin of the thus enhanced manetic moment in the FeCo system has been discussed based on the scattered-wave X_α-cluster calculation [7.79]. Also the electronic band structure of ordered prototype systems FeCo, Fe_3Co and $FeCo_3$ has been calculated by the augmented spherical waves method [7.80]. According to the second study [7.80], two important processes account for the unusually high magnetization of the FeCo system. First, the spin-up states at the Fe site are completely filled due to additional electrons provided by Co atoms, so that the alloy becomes a strong ferromagnet. Second, the further filling of spin-down states by these additional electrons is prevented by the fact that it would result in an increase in the density of states at the Fermi level, and it is therefore energetically unfavourable. In other words, the Fermi level is pinned in a pronounced dip in the density of spin-down states.

These general observations concerning the DOS mostly remain valid as substitutional $3d$ impurities are added to the FeCo alloy [7.78]. Figure 7.26 shows these DOS, and Table 7.10 lists the calculated values of magnetic moments and exchange integrals.

Despite the fact that impurities entering the Co site of the FeCo alloy have the same nearest neighbours as impurities in pure bcc Fe, the evolution of the magnetic moment in the row of impurities differs considerably for these two systems (Fig. 7.27). An inversion of the direction of the moment occurs between

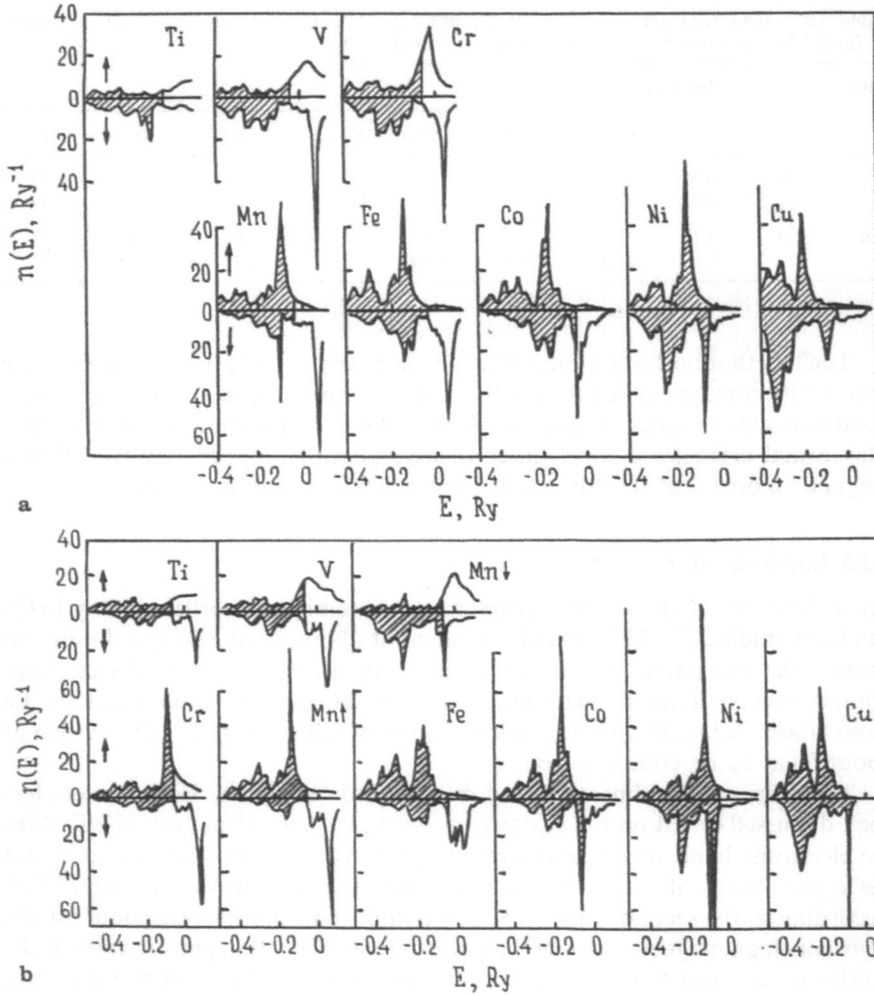

Fig. 7.26. Local DOS of 3*d* impurities substituted into the (**a**) Co and (**b**) Fe sites in the FeCo alloy

Table 7.10. Calculated values of local magnetic moments $M[\mu_B]$ and effective exchange parameters J_0[m Ry] for 3*d* impurities in FeCo alloy

Substitution site		Impurities								
		Ti	V	Cr	Mn ↓	Mn ↑	Fe	Co	Ni	Cu
Fe	$M[\mu_B]$	− 0.38	− 0.14	1.88	− 3.05	2.94	2.66	1.41	0.01	− 0.02
	J_0[m Ry]	0.9	0.1	9.4	− 27.5	16.1	17.6	7.4	− 0.6	− 0.2
Co	$M[\mu_B]$	− 0.65	− 0.94	− 0.39	—	2.39	2.36	1.72	0.33	0.09
	J_0[m Ry]	3.7	0.5	0.8	—	14.7	21.1	17.8	0.7	− 0.3

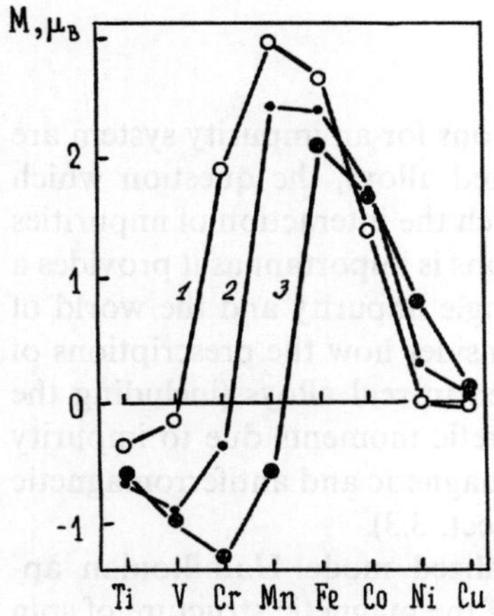

Fig. 7.27. Local magnetic moments of 3*d* impurities substituted into the Fe (*1*) and Co (*2*) sites in FeCo, and of 3d impurities in pure bcc Fe (*3*)

Cr and Mn, so that for Mn only "parallel" orientation of the magnetic moment is found. Keeping in mind the results discussed in pevious sections, this is a very unusual situation for Mn. Figure 7.26 shows that the orign of the large magnetic moment of the Mn impurity is in a reduced population of spin-down states due to the above-mentioned pinning of the Fermi level in a state density dip.

The magnetic moments of impurities substituted into the Fe site in FeCo follow the same trends in the row of impurities (Fig. 7.27), but now the inversion of the magnetic moment occurs even for the Cr impurity. Two self-consistent solutions have been found for the Mn impurity, but the one with the "reversed" moment is very unstable, as follows from the analysis of J_0 (Table 7.10). The remaining solution for the Mn impurity in the Fe site corresponds to the highest value among all FeCo-based impurity systems.

Note that this large magnetic moment, although partly compensated by the polarization of the opposite sign in the surrounding crystal, is not completely cancelled, so replacement of the Fe atom by Mn in FeCo leads to a net gain of magnetization in the crystal. This gain equals $\simeq 0.18\mu_B$ per Mn atom, as has been estimated [7.78] from Lloyd's formula for the induced magnetic moment. Replacement of the Co by Mn results in no loss of total magnetic moment. Therefore, doping the FeCo alloy with Mn would result in a slight increase of the net magnetization. This is consistent with the result of band-structure calculations of the fictitious Fe_2CoMn and Co_2FeMn intermetallics discussed in Chap. 3. According to these calculations, even a Mn content as high as 25 at. % still gives an increase of magnetization over that for FeCo.

A peculiar feature of impurity DOS for some 3*d* impurities in FeCo is the appearance of narrow peaks in the region of reduced density of host states near the Fermi level. These quasi-discrete non-bonding states are of the same origin as those reported for impurities in Cr and Mn near the "antiferromagnetic dip". These states are situated ~ 0.03 Ry below the Fermi level; it seems possible that they could be detected by some spectroscopic techniques.

7.4 Interaction of Impurities

When the results of electronic structure calculations for an impurity system are applied to the discussion of properties of diluted alloys, the question which usually remains open is the concentration to which the interaction of impurities can be neglected. The study of impurity interactions is important as it provides a bridge between the rather idealized case of a single impurity and the world of alloy systems. So, it is of practical interest to consider how the prescriptions of the Alexander–Andersen model are implemented in real alloys (including the suppression of the tendency to form local magnetic moments due to impurity interactions, and the competition between ferromagnetic and antiferromagnetic interactions between magnetic moments – see Sect. 3.3).

Deeper insight than that provided by idealized model-Hamiltonian approaches is necessary for better understanding of the magnetic structure of spin glasses and micromagnets, like CuMn or AuFe [7.81, 82].

Oswald et al. have calculated the electronic structure of interacting *3d* impurities in Cu and Ag [7.83] by the KKR–Green function method; the results are also discussed in [7.27, 84]. The reduced point group symmetry was considered in order to account for both ferromagnetic and antiferromagnetic couplings within the impurity pair.

Figures 7.28, 29 show the density of states of interacting Mn impurities in Cu and Ag. The DOS for a single impurity is shown for comparison in both cases as a dotted line. For ferromagnetically coupled impurity moments, the splitting of the virtual bound-state peak is quite pronounced in the spin-down DOS and, to a lesser extent, in the spin-up DOS. The difference is due to the fact that spin-up impurity states are more localized and hence less overlapping than the spin-down states. The splitting is not a mere splitting into two similar peaks as it was in the simple Alexander–Anderson model (Sect. 3.3). Instead, a single Lorentzian-shaped peak decomposes into several asymmetric contributions

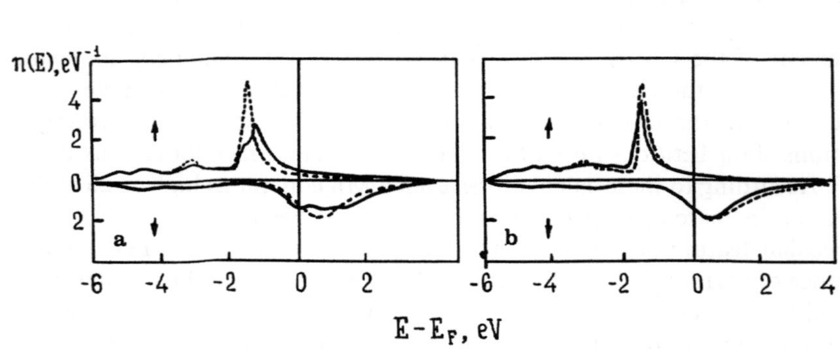

Fig. 7.28. Local DOS of neighbouring Mn impurities in Cu for cases of (**a**) ferromagnetic and (**b**) antiferromagnetic configurations. Local DOS of a single Mn impurity is shown as a dashed line

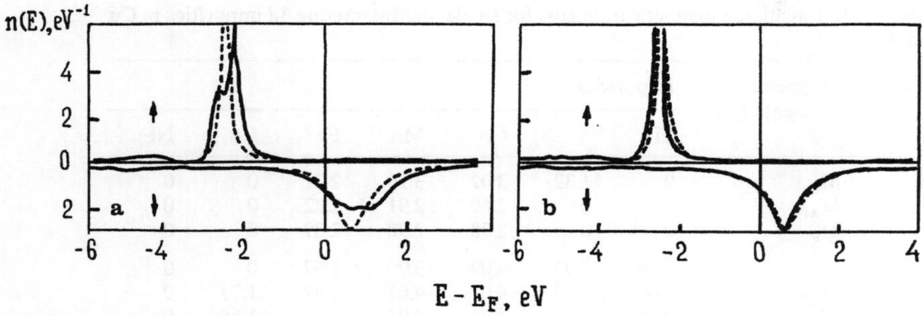

Fig. 7.29a, b. Same as Fig. 7.28, but for Mn impurities in Ag

from several hybrid states according to the reduced point group symmetry [7.83].

With antiferromagnetic coupling of magnetic moments, the difference of the DOS from that of a single impurity is much less pronounced. It is restricted, in fact, to the appearance of small humps at the energies corresponding to the position of a peak for the opposite spin direction. It reveals the interaction with electrons of the same spin at another impurity site.

All these comparatively small effects of interaction are more important for impurities in Cu than for those in Ag. Because the lattice constant of Ag is 15% larger than that of Cu, the overlapping of impurity *d* states is correspondingly reduced.

Table 7.11 lists the values of local magnetic moments of single and interacting impurities. Note that the magnetic moment of the V impurity in Cu disappears when the interaction with another similar impurity is included. For Ti in Ag, their interaction reduces the magnetic moment by 30% [7.27]. This can be understood on the basis of the above-discussed criterion for the existence of a magnetic moment (Sect. 3.5), according to which, a high density of impurity states at E_F in the non-spin-polarized case favours spin splitting. Even if this criterion is satisfied for a single impurity whose virtual bound state crosses the Fermi level, the splitting of this impurity state due to interaction with that of another impurity lowers the DOS in some cases below the critical level.

The important question of the energy preference of one or another type of moment orientation has been considered [7.27, 83]. Once the binding energies for impurities in antiferromagnetic (AF) or ferromagnetic (F) configurations are calculated, not only the most stable one can be found, but also the exchange interaction energy for the impurity pair can be estimated.

The term "binding energy" as used by *Oswald* et al. [7.83] means the difference of the total energies of two systems: the interacting pair of impurities in the host, the magnetic moments of which are aligned parallel or antiparallel, and the corresponding pair of impurities, the interaction between which is "switched off". The energy of the latter system is actually double the total energy

Table 7.11. Calculated magnetic moments for single and interacting 3*d* impurities in Cu and Ag

Host	Magnetic moments [μ_B]	Impurities						
		Ti	V	Cr	Mn	Fe	Co	Ni
Cu	M_0	0	1.32	3.02	3.16	2.21	0	0
	M_{AF}	0	0	2.60	2.91	2.22	0	0
	M_F	0	0	2.78	2.98	2.09	0	0
Ag	M_0	1.40	2.93	4.09	3.00	1.67	0	0
	M_{AF}	1.00	2.82	4.07	4.01	2.97	1.73	0
	M_F	0.81	2.74	4.01	4.04	2.97	1.68	0

M_0 is the local magnetic moment of a single impurity;
M_F and M_{AF} are local magnetic moments in ferro- and antiferromagnetic impurity pairs

of a single impurity, and the mutual orientation of magnetic moments of impurities does not affect this property. Each of these total energies includes a single-particle part and double-counting terms. Their direct calculation would be troublesome, and it would result in numerical errors in determining the binding energies as differences of two large numbers. It has been pointed out [7.83] that the double-counting energy is almost exactly cancelled in the final difference of energies, provided they are calculated with the same potential. The error introduced is second order in magnetization, and it can be neglected because the magnetic-moment value is usually only slightly affected by inter-impurity interaction. The described procedure supposes that the effect of interaction is considered only in the one-electron part via the local densities of states, which are different for interacting and non-interacting impurities even if calculated with the same potential.

The resulting formula for the binding energy used by *Oswald* et al. [7.83] is as follows:

$$E_B = \int^{E_F} dE(E_F - E)[n_{1(2)}(E) + n_{2(1)}(E) - n_1(E) - n_2(E)], \tag{7.5}$$

where $n_1(E)$ is the local DOS of a single impurity 1, and $n_{1(2)}(E)$ is the local DOS of impurity 1 in the presence of impurity 2. To obtain (7.5), *Oswald* et al. used the additional simplifying assumption that the spatial magnetization is restricted to the impurity atomic sphere; i.e. the host polarization is neglected. However, they pointed out [7.83], the binding energy can be expressed in a similar way in terms of integrated densities of states. Table 7.12 gives the results of the calculation of binding energies for ferromagnetic and antiferromagnetic impurity pairs in Cu and Ag. Cases where impurities are the nearest and more distant (up to the 4th shell) neighbours are considered. From the differences of binding energies for two magnetic configurations of impurities (F and AF), the RKKY exchange parameters have been estimated by the formula

$$J = \frac{E_B(F) - E_B(AF)}{2S(S + 1/2)}, \tag{7.6}$$

Table 7.12. Magnetic interaction energies for impurities in Cu and Ag

Energies	Impurities in Cu			Impurities in Ag				
	Cr	Mn	Fe	V	Cr	Mn	Fe	Co
E^1_F	−116	76	225	−9	−196	−37	108	−13
E^1_{AF}	248	161	66	115	19	−7	−70	−143
ΔE_1	364	−85	159	−124	−215	−30	178	130
J_1	−696	−150	520	−250	−240	−32	344	677
E^2_F	46	44	−14	6	23	35	−1	−16
E^2_{AF}	4	−11	−8	37	13	5	4	5
ΔE_2	42	55	−6	−31	10	30	−5	−21
J_2	80	97	−20	−62	11	33	−10	−109
E^2_F	−5	−14	10	−11	−3	10	15	−4
E^3_{AF}	5	−8	−13	16	0	−2	−2	−4
ΔE_3	−16	22	23	−27	−3	12	17	0
J_3	−19	39	75	−54	−3	13	33	0
E^4_F	−2	−21	8	3	0	−14	−25	−2
E^4_{AF}	−18	1	−20	−32	−8	−1	1	2
ΔE_4	16	−22	−28	35	8	−13	−26	−4
J_4	31	−39	−92	71	9	−15	−50	−21

$E^n_{F(AF)}$ are bonding energies (in meV) of ferromagnetic (antiferromagnetic) impurity pairs which are nth nearest neighbours;
$E_n = E^n_F - E^n_{AF}$;
$J_n = (\Delta E_n)/[2S(S+1/2)]$ is the RKKY exchange parameter (in K; S is the spin of the impurity atom; $\mu_B S = 1/2 M_0$)

where $S = M_0/2\mu_B$. The values of these exchange parameters are also given in the Table 7.12. They made it possible to discuss the spatial variation of the RKKY interaction in Cu- and Ag-based alloys.

As Table 7.12 shows, the binding energies for all nearest-neighbour impurity pairs in Cu, with the exception of ferromagnetically coupled Cr impurities, are positive. On the other hand, negative binding energies are more typical for impurities in Ag. Because the direct exchange interaction of d states always favours binding, one can conclude that it is dominant in Cu-based alloys, while in Ag-based alloys, the indirect exchange interaction plays an important role.

The binding energy for the AF configuration of Cr impurities in Cu is larger than that of Mn impurities. Most probably this is due to the fact that the d states become more localized along the $3d$ row.

The maximum binding energy among all ferromagnetically coupled impurities is for the Fe pair. This makes sense because the virtual bound-state peak for spin-down states is situated just at the Fermi energy, and its splitting due to impurity interaction lowers the DOS at E_F and hence the total energy. The even higher binding energy value for an antiferromagnetically coupled pair of Cr impurities in Cu reveals the fact well-known from the Alexander–Anderson model (Sect. 3.3) that the impurities with a half-filled d shell tend to antiparallel coupling of their magnetic moments.

Experimental data concerning the magnetic interaction of Mn impurities in Cu are as follows. It was traditionally believed that the Mn are coupled antiferromagnetically, but neutron scattering measurements [7.85, 86] indicated the presence of ferromagnetic coupling. Further studies [7.87] have revealed that pairs of Mn atoms are antiferromagnetically coupled when they are the nearest neighbours, but ferromagnetic coupling occurs for next-nearest neighbours. This agrees with calculational results [7.83]. To our knowledge, corresponding experimental data on neutron scattering in the CuCr alloy are not available.

Exchange interactions dependent on the interimpurity interaction in CuCr and CuFe were theoretically studied within the tight-binding approach [7.88]. Overall, these model calculations agree with ab initio results [7.83].

Measurements of high-temperature spin-glass susceptibilities [7.89] permitted extraction of the values of local exchange integrals for some Cu- and Ag-based alloys. The signs of the RKKY exchange parameters calculated by *Morgownik* and *Mydosh* [7.89] differ from those reported by *Oswald* et al. [7.83] for the first and third neighbours of Mn in Cu and Mn in Au, and for third neighbours of Fe in Au. Note, however, that the determination of the exchange integrals in the first study [7.89] was not straightforward and used the results of cluster calculations based on the Ising Hamiltonian. At least in one point – namely, the sign of exchange interaction between Mn atoms which are the nearest neighbours in Cu – the results [7.89] contradict the neutron scattering data [7.87].

The method used by *Oswald* et al. [7.83] to evaluate the exchange interaction energy is only asymptotically correct, and for considerable magnitudes of the magnetic moment the error introduced is not negligible. As the authors pointed out [7.83], the exchange energy estimates, at least for Mn in Cu and in Ag, are rather sensitive to the trial fixed potential used in the calculation (i.e. if the potential is self-consistent for a single impurity, or self-consistent for the ferro- or antiferromagnetic impurity pair). In other words, the error in the estimate of the total energy of the ferromagnetic impurity pair with the non-self-consistent potential used is too large and not completely cancelled by the corresponding error in the antiferromagnetic pair calculation.

This is mostly true in the special case of Mn impurities because they are a critical border region near the transition from predominantly antiferromagnetic to predominantly ferromagnetic coupling within an impurity pair. Therefore, the calculational results for these impurities are very sensitive to small variations of the potential.

Another way to treat magnetic interactions, not via the direct calculation of bonding energies, has been proposed in [7.90]. Its basis is the introduction of constraints on the magnetic-moment values. It is clear that any a priori chosen value of the magnetic moment M at an impurity site can be stabilized by introducing an additional magnetic field $H(M)$. According to the Hellmann–Feynman theorem, this magnetic field is $H = \partial E(M)/\partial M$. Then, if the $H(M)$ dependence as a whole is known, the total energy for any magnetic configura-

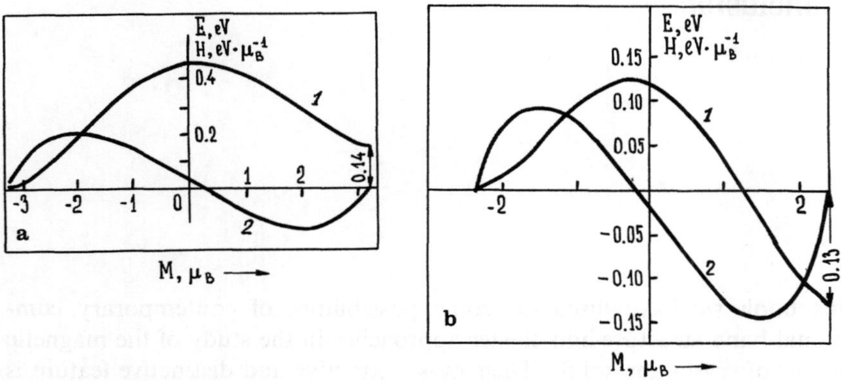

Fig. 7.30. Totral energy E relative to that of the antiferromagnetic configuration (*1*) and effective magnetic field (*2*) as functions of the magnetic moment of the second impurity for (**a**) Mn and (**b**) Fe impurity pairs in Cu

tion can be calculated. *Oswald* et al. [7.90] have implemented this scheme as follows. Starting from the self-consistent antiferromagnetic configuration chosen as a reference point, the magnetic moment of a second impurity is arbitrarily changed from a negative value M_0 to zero and then to positive values, as the magnetic field $H(M)$ is introduced at this impurity site. The potential of the first impurity is kept fixed. Then, the energy difference with respect to the energy of the antiferromagnetic configuration is

$$\Delta E(M) = E(M) - E_{\mathrm{AF}} = \int_{M_0}^{M} H(M')\,dM'. \tag{7.7}$$

One has to see whether there are local minima of this function in addition to that at $M = M_0$. The energy gain or loss compared with that of the antiferromagnetic configuration is found immediately. Figure 7.30 shows two typical cases of the $\Delta E(M)$ behaviour for Mn and Fe impurity pairs in Cu. For CuMn, the initial antiferromagnetic configuration has energy lower by 0.14 eV than that of the ferromagnetic configuration, while for CuFe the ferromagnetic configuration lies 0.13 eV lower in energy. The calculation [7.90] considerably corrected the earlier results [7.83] concerning Mn in Cu, using a more accurate method to estimate the energy differences. Moreover, the KKR–Green function calculations were performed [7.90] for the embedded cluster including impurities and their nearest neighbours, while earlier [7.83] only the potential in the impurity spheres was allowed to change during the self-consistent procedure.

Another way of changing the magnetic moment in the calculation scheme described above is through deviation of the moment from its original collinear configuration, with the moment direction being introduced as a constraint. As has been mentioned [7.90], the interaction energy in this case follows the $\sim \cos\theta$ dependence typical for the usual Heisenberg model.

8. Conclusion

In this book we have discussed some possibilities of contemporary computational band structure and cluster approaches in the study of the magnetic properties of crystalline solids. Their most attractive and distinctive feature is that, unlike phenomenological models, they enable one to take complete account of the electronic structure and chemical bonding peculiarities of the crystal under consideration. When describing the magnetic properties of solids, such an approach, as a rule, makes unnecessary the use of any phenomenological, and sometimes not well defined parameters, which depend on electronic structure features and are typical for qualitative models of magnetic interactions in solids. As we have shown, investigation of relatively weak interactions of magnetic moments in crystals in such an approach can be closely connected with direct calculations of their electronic structure, process of magnetic-moment formation and spin excitations, which define magnetic interactions in solids.

We have considered crystals with different types of lattice, magnetic ordering and degree of localization of magnetic electron states. As is obvious from the results presented, contemporary approaches developed in the framework of predominantly local spin-density functional theory usually provide a relatively simple and unique way to study peculiarities of the magnetic structures of crystals, to find and explain the most stable type of magnetic ordering, and, for every system under consideration, to calculate magnetic excitation energies and critical temperatures for magnetic transitions with sufficient accuracy.

The accuracy of the calculations discussed for complicated many-electron systems appears impressive and allows one to evaluate exchange integrals of the order of several degrees Kelvin, even if they are defined by differences of large total energies of the systems, which are of the order of 10^5 eV and more.

The applications of calculations based on the spin-density functional formalism to the study of magnetic systems have become quite numerous now and include the treatment of metals, alloys and compounds of *d* and *f* elements with various crystal and magnetic structures. We hope that this book will encourage readers to think about using the methods described in their investigations of magnetic systems and interpretation of experimental results.

References

Chapter 1

1.1 S.V. Vonsovskii: *Magnetism*, Vols. 1, 2 (Wiley, New York 1974)
1.2 C.M. Hurd: Contemp. Phys. **23**, 469–493 (1982)
1.3 Y.A. Izyumov, R.P. Ozerov: *Magnetic Neutron Diffraction* (Plenum, New York 1970)
1.4 R.M. Wite, T.H. Geballe: Long range order in solids, in *Solid State Physics*, Suppl. 15, ed. by H. Ehrenreich, F. Seitz, D. Turnbull (Academic, New York 1979)
1.5 T. Moriya: Phys. Rev. **120**, 91–98 (1960)
1.6 W. Heisenberg: Z. Phys. **49**, 619–631 (1928)
1.7 J. Goodenough: Magnetism and crystal structure in nonmetals, in *Magnetism*, Vol. 3, ed. by G.T. Rado, H. Suhl (Academic, New York 1966) p. 1
1.8 P.A.M. Dirac: Proc. R. Soc. London A **123**, 714–728 (1929)
1.9 J.H. van Vleck: *The Theory of Electronic and Magnetic Susceptibilities* (Oxford University Press, Oxford 1932)
1.10 S.V. Tyablikov: *Methods in the Quantum Theory of Magnetism* (Plenum, New York 1967)
1.11 J. Smart: Evaluation of exchange interactions from experimental data, in *Magnetism*, Vol. 2, ed. by G.T. Rado, H. Suhl (Academic, New York 1963) p. 63
1.12 A.J. Freeman, R.E. Watson: Phys. Rev. **124**, 1439–1454 (1961)
1.13 J.I. Frenkel: Z. Phys. **49**, 31–42 (1928)
1.14 F. Bloch: Z. Phys. **57**, 545–558 (1929)
1.15 E.C. Stoner: Proc. R. Soc. London A **165**, 372–381 (1938)
1.16 C. Herring: Exchange interactions among itinerant electrons, in *Magnetism*, Vol. 4, ed. by G.T. Rado, H. Suhl (Academic, New York 1964) pp. 1–407
1.17 E. Wigner: Phys. Rev. **46**, 1002–1019 (1934)
1.18 D.M. Ceperley, B.J. Adler: Phys. Rev. Lett. **45**, 566–569 (1980)
1.19 D.R. Salahub, R.P. Messmer: Surf. Sci. **106**, 415–421 (1981)
1.20 T. Moriya: *Spin Fluctuations in Itinerant Electron Magnetism*, Springer Ser. Solid-State Sci., Vol. 56 (Springer, Berlin, Heidelberg 1985)
1.21 P.W. Anderson: Exchange in insulators: Superexchange, direct exchange and double exchange, in *Magnetism*, Vol. 1, ed. by G.T. Rado, H. Suhl (Academic, New York 1964) pp. 25–84

Chapter 2

2.1 H.A. Kramers: Physica **1**, 182–187 (1934)
2.2 P.W. Anderson: Phys. Rev. **79**, 350–358 (1950)
2.3 F. Keffer, T. Oguchi: Phys. Rev. **115**, 1428–1434 (1959)
2.4 N.L. Huang, R. Orbach: Phys. Rev. **154**, 487–492 (1967)
2.5 P.W. Anderson: Theory of magnetic exchange interaction: exchange in insulators and semiconductors, in *Solid State Physics*, Vol. 14, ed. by F. Seitz, D. Turnbull (Academic, New York 1963) pp. 99–214
2.6 G.W. Pratt: Phys. Rev. **97**, 926–932 (1955)
2.7 P.W. Anderson: Phys. Rev. **115**, 2–13 (1959)
2.8 J.C. Slater: *The Self-Consistent Field for Molecules and Solids* (McGraw-Hill, New York 1974)
2.9 B.H. Brandow: Adv. Phys. **26**, 651–808 (1977)
2.10 K.I. Gendaira, Y. Tanabe: J. Phys. Soc. Jpn. **21**, 1527–1548 (1966)

2.11 R.K. Nesbet: Phys. Rev. **122**, 1467–1478 (1961)
2.12 S. Freeman: Phys. Rev. B **7**, 3960–3986 (1973)
2.13 N. Fuchikami: J. Phys. Soc. Jpn. **28**, 871–887 (1970)
2.14 G. van Kalkeren, W.W. Schmidt, R. Block: Physica B + C **97**, 315–337 (1979)
2.15 N. Fuchikami, Y. Tanabe: J. Phys. Soc. Jpn. **45**, 1559–1566 (1978)
2.16 N. Fuchikami, Y. Tanabe: J. Phys. Soc. Jpn. **47**, 505–511 (1979)
2.17 K.W.H. Stevens: Phys. Rep. **24**, 1–75 (1976)
2.18 A.A. Sidorov, A.S. Moskvin, V.V. Popkov: Fiz. Tverd. Tela **18**, 3005–3013 (1976) [Engl. transl.: Sov. Phys. – Solid State. **18**, 1751–1755 (1976)]
2.19 N.N. Bogolubov: *Lecture on Quantum Statistics* Vol. 1 (Gordon and Breach, New York 1970)
2.20 J. Slater: Phys. Rev. **165**, 655–658 (1968)
2.21 M.V. Eremin: Fiz. Tverd. Tela **24**, 423–432 (1982) [Engl. transl.: Sov. Phys. – Solid State **24**, 239–244 (1982)]
2.22 M.V. Eremin: Fiz. Tverd. Tela **24**, 3216–3222 (1983) [Engl. transl.: Sov. Phys. – Solid State **25**, 1010–1013 (1983)]
2.23 M.V. Eremin, Yu.V. Rakitin: J. Phys. C **15**, L259–L261 (1982)
2.24 B. Koiller, L.M. Falikov: J. Phys. C **8**, 695–707 (1975)
2.25 I. Mayer, S.A. Angelov: J. Phys. C **16**, L857–L858 (1983)
2.26 M.V. Eremin, Yu.V. Rakitin: Phys. Status Solidi (b) **80**, 579–587 (1977)
2.27 E.L. Nagaev: Usp. Fiz. Nauk **136**, 61–103 (1982) [Engl. transl.: Sov. Phys. – Uspekhi **25**, 31̄57 (1982)]
2.28 T. Moriya: Phys. Rev. **120**, 91–98 (1960)
2.29 E.L. Nagaev: Zh. Eksp. Teor. Fiz. **74**, 1375–1385 (1978) [Engl. transl.: Sov. Phys. – JETP **47**, 719–725 (1978)]
2.30 M.I. Auslender, M.I. Katsnelson· Teor. Mat. Fiz. **51**, 436–444 (1982)
2.31 K.I. Kugel, Khomskii D.I.: Usp. Fiz. Nauk **136**, 621–664 (1982) [Engl. transl.: Sov. Phys. – Uspekhi **25**, 231–256 (1982)]
2.32 K.I. Kugel, Khomskii D.I.: Zh. Eksp. Teor. Fiz. **64**, 1429–1436 (1973) [Engl. transl.: Sov. Phys. – JETP **37**, 725–730 (1973)]
2.33 K.I. Kugel, Khomskii D.I.: Fiz. Tverd. Tela **17**, 454–461 (1975) [Engl. transl.: Sov. Phys. – Solid State **17**, 285–289 (1976)]
2.34 K. Hirakawa, Y. Kurogi: Prog. Theor. Phys. **46S**, 147–161 (1975)
2.35 M.V. Eremin, V.P. Kalinenkov: Fiz. Tverd. Tela **23**, 1422–1427 (1981) [Engl. transl.: Sov. Phys. – Solid State **23**, 828–831 (1981)]
2.36 J. Zaanen: The electronic structure of transition metal compounds in the impurity model. Ph.D. Thesis, University of Groningen (1986)
2.37 N.F. Mott: Proc. Phys. Soc. (London) A **62**, 416–421 (1949)
2.38 J. Hubbard: Proc. R. Soc. London A **277**, 237 (1964)
2.39 S. Hüfner: Solid State Commun. **49**, 1177–1179 (1984)
2.40 J. Pollini, J. Thomas, G. Jezequel, J.C. Lemonnier, A. Lenselik: Phys. Rev. B **29**, 4716 (1984)
2.41 D. Adler, J. Feinleib: Phys. Rev. B **2**, 3112–3134 (1970)
2.42 O. Gunnarsson, K. Schoenhammer: Phys. Rev. Lett. **50**, 604–611 (1983)
2.43 J. Zaanen, G.A. Sawatzky, J.W. Allen: Phys. Rev. Lett. **55**, 418–421 (1985)
2.44 J. Zaanen, G.A. Sawatzky: Can. J. Phys. **65**, 1262–1271 (1987)

Chapter 3

3.1 J. Friedel: Can. J. Phys. **34**, 1190–1211 (1956)
3.2 J. Friedel: Nuovo Cimento, Ser. X, **VII**, suppl., 287–311 (1958)
3.3 J. Friedel: J. de Phys. **19**, 573–581 (1958)
3.4 P.W. Anderson: Phys. Rev. **124**, 41–53 (1961)
3.5 S. Alexander, P.W. Anderson: Phys. Rev. **133**, A1594–A1603 (1964)
3.6 T. Moriya: Prog. Theor. Phys. **33**, 157–183 (1965)
3.7 L. Hirst: Phys. kondens. Mater. **11**, 255–278 (1970)
3.8 O. Gunnarsson: J. Phys. F **6**, 587–606 (1976)
3.9 J.K. Janak: Phys. Rev. B **16**, 255–262 (1977)
3.10 K. Terakura, N. Hamada, T. Oguchi, T. Asada: J. Phys. F **12**, 1661–1678
3.11 V. Heine, J.H. Samson, C.M.M. Nex: J. Phys. F **11**, 2645–2662 (1981)

Chapter 4

4.1 S.V. Yonsovsky: *Magnetism*, Vols. 1, 2 (Wiley, New York 1974)
4.2 T. Moriya (ed.): *Electron Correlation and Magnetism in Narrow-Band Systems*, Springer Ser. Solid-State Sci., Vol. 29 (Springer, Berlin, Heidelberg 1981)
4.3 J. Hubbard: Proc. R. Soc. London A **276**, 238–257 (1963)
4.4 F. Gautier: Itinerant magnetism, in *Magnetism of Metals and Alloys*, ed. by M. Cyrot (North-Holland, Amsterdam 1982)
4.5 E.C. Stoner: Proc. R. Soc. London A **165**, 372–381 (1938)
4.6 R.M. White: *Quantum Theory of Magnetism* (McGraw-Hill, New York 1970)
4.7 C. Herring: Exchange interactions among itinerant electrons, in *Magnetism*, ed. by G.T. Rado, H. Suhl, Vol. 4 (Academic, New York 1966) pp. 1–407
4.8 T. Moriya: J. Magn. Magn. Mater. **14**, 1–46 (1979)
4.9 T. Moriya: J. Magn. Magn. Mater. **31/34**, 10–18 (1983)
4.10 J. Hubbard: Phys. Rev. B **19**, 2626–2636 (1979)
4.11 Yu.P. Irkhin, E.W. Rosenfeld: Solid State Commun. **44**, 1371–1373 (1982)
4.12 K.K. Murata, S. Doniah: Phys. Rev. Lett. **29**, 285–288 (1972)
4.13 T. Moriya, A. Kawabata: J. Phys. Soc. Jpn. **34**, 639–651 (1973)
4.14 R.L. Stratonovich: Dok. Akad. Nauk SSSR **157**, 1097–1100 (1957) [Engl. transl.: Sov. Phys. – Dokl. **2**, 416–419 (1958)]
4.15 J. Hubbard: Phys. Rev. Lett. **3**, 77–78 (1959)
4.16 S.Q. Wang, W.E. Evenson, J.R. Schrieffer: Phys. Rev. Lett. **23**, 92–95 (1969)
4.17 M. Cyrot: Phys. Lett. A **37**, 189–190 (1971)
4.18 M.V. You, V. Heine, A.J. Holden, P.J. Lin-Chang: Phys. Rev. Lett. **44**, 1282–1284 (1980)
4.19 J. Hubbard: Phys. Rev. B **20**, 4584–4595 (1979)
4.20 P. Lacour-Gayet, M. Cyrot: J. Phys. C **7**, 400–408 (1974)
4.21 H. Cappelman: J. Phys. F **4**, 1112–1120 (1974)
4.22 H. Hasegawa: J. Phys. Soc. Jpn **46**, 1504–1514 (1979)
4.23 H. Hasegawa: J. Phys. F **13**, 1915–1929 (1983)
4.24 V.I. Grebennikov, Yu.I. Prokopiev, O.B. Sokolov, E.A. Turov: Fiz. Met. Metalloved. **52**, 679–694 (1981)
4.25 J. Hubbard: J. Appl. Phys. **52**, 1654–1657 (1981)
4.26 H. Hasegawa: J. Phys. Soc. Jpn. **49**, 178–188 (1979)
4.27 V. Korenman, J.L. Murray, R.E. Prange: Phys. Rev. B **16**, 4032–4047 (1977)
4.28 V. Korenman, J.L. Murray, R.E. Prange: Phys. Rev. B **16**, 4048–4057 (1977)
4.29 V. Korenman, J.L. Murray, R.E. Prange: Phys. Rev. B **16**, 4058–4067 (1977)
4.30 R.E. Prange, V. Korenman: Phys. Rev. B **19**, 4691–4697 (1979)
4.31 R.E. Prange, V. Korenman: Phys. Rev. B **19**, 4698–4702 (1979)
4.32 H. Capellmann: Z. Phys. B **34**, 29–35 (1979)
4.33 D.M. Edwards: J. Magn. Magn. Mater. **36**, 213–216 (1983)
4.34 P.J. Brown, H. Capellmann, J. Deportes et al.: J. Magn. Magn. Mater. **36**, 335–339 (1983)
4.35 A. Ziegler: Phys. Rev. Lett. **48**, 695–698 (1982)
4.36 T. Usami, T. Moriya: J. Magn. Magn. Mater. **20**, 171–194 (1980)
4.37 T. Moriya: *Spin Fluctuations in Itinerant Electron Magnetism*, Springer Ser. Solid-State Sci., Vol. 56 (Springer, Berlin, Heidelberg 1985)
4.38 M.V. You, V. Heine: J. Phys. F **12**, 177–194 (1982)
4.39 R. Haydock, V. Heine, M.J. Kelly: J. Phys. C **8**, 2591–2605 (1975)
4.40 V. Heine: Electronic structure from the point of view of the local atomic environment, in *Solid State Physics*, Vol. 35 ed. by H. Ehrenreich et al. (Academic, New York 1980) pp. 1–127
4.41 A.J. Holden, M.V. You: J. Phys. F **12**, 195 (1982)
4.42 P.J. Lin-Chang, A.J. Holden: Phys. Rev. B **23**, 3414–3420 (1981)
4.43 J.W. Linn: Phys. Rev. B **11**, 2624 (1975)
4.44 L.M. Sandratskii: Phys. status solidi (b) **136**, 167 (1986)
4.45 L.M. Sandratskii, P.G. Guletskii: J. Phys. F **16**, L43 (1986)
4.46 L.M. Sandratskii, P.G. Guletskii: J. Magn. Magn. Mater. **79**, 306–320 (1989)
4.47 L.M. Sandratskii, P.G. Guletskii: Phys. status solidi (b) **154**, 623–636 (1989)

Chapter 5

5.1 D.R. Hartree: Proc. Cambridge Philos. Soc. **24** 89, 111, 426 (1928)
5.2 V. Fock: Z. Phys. **61**, 126–148 (1930)
5.3 P. Hohenberg, W. Kohn: Phys. Rev. **136**, B864–B872 (1964)
5.4 O. Gunnarsson, M. Jonson, B.I. Lundqvist: Phys. Rev. B **20**, 3136–3164 (1979)
5.5 R. Gaspar: Acta Phys. Acad. Sci. Hung. **24**, 173 (1968)
5.6 W. Kohn, L.J. Sham: Phys. Rev. **140**, A1133–A1138 (1965)
5.7 J.C. Slater: *The Self-Consistent Field for Molecules and Solids* (McGraw-Hill, New York 1974)
5.8 K. Schwarz: Phys. Rev. B **5**, 2466–2468 (1972)
5.9 U. von Barth, L. Hedin: J. Phys. C **5**, 1629–1642 (1972)
5.10 L. Hedin, B.I. Lundqvist: J. Phys. C **4**, 2064–2083 (1971)
5.11 O. Gunnarsson, B.I. Lundqvist: Phys. Rev. B **13**, 4274–4298 (1976)
5.12 V.L. Moruzzi, J.F. Janak, A.R. Williams: *Calculated Electronic Properties of Metals* (Pergamon, New York 1978)
5.13 D.M. Ceperley, B.J. Alder: Phys. Rev. Lett. **45**, 566–569 (1980)
5.14 S.H. Vosko, L. Wilk, M. Nusair: Can. J. Phys. **58**, 1200–1211 (1980)
5.15 P. Bagno, O. Jepsen, O. Gunnarsson: Phys. Rev. B **40**, 1997–2000 (1989)
5.16 D.C. Langreth, M.J. Mehl: Phys. Rev. Lett. **47**, 446–450 (1981)
5.17 D.C. Langreth, M.J. Mehl: Phys. Rev. B **28**, 1809–1834 (1983)
5.18 C.D. Hu, D.C. Langreth: Phys. Rev. B **33**, 943–959 (1986)
5.19 J.P. Perdew: Phys. Rev. B **33**, 8822–8824 (1986)
5.20 J.P. Perdew: Phys. Rev. B **34**, 7407–7412 (1986)
5.21 J.P. Perdew, Y. Wang: Phys. Rev. B **33**, 8800–8802 (1986)
5.22 J.P. Perdew, A. Zunger: Phys. Rev. B **23**, 5048–5079 (1981)
5.23 R.O. Jones, O. Gunnarsson: Rev. Mod. Phys. **61**, 689–746 (1989)
5.24 R. Podloucky, R. Zeller, P.H. Dederichs: Phys. Rev. B **22**, 5777–5790 (1980)
5.25 B.L. Gyorffy, M.J. Stott: A one-electron theory of soft X-ray emission from random alloys, in *Band Structure Spectroscopy of Metals and Alloys* (Academic, New York 1973) pp. 385–403
5.26 G. Ries, H. Winter: J. Phys. F **9**, 1589–1609 (1979)
5.27 J. Korringa: Physica **13**, 392–400 (1947)
5.28 W. Kohn, N. Rostoker: Phys. Rev. **94**, 1111–1120 (1954)
5.29 P.J. Braspening, R. Zeller, A. Lodder, P.H. Dederichs: Phys. Rev. B **22**, 703–718 (1984)
5.30 O.K. Andersen: Phys. Rev. B **8**, 3060–3083 (1975)
5.31 C. Koenig, N. Stefanou, J.M. Koch: Phys. Rev. B **33**, 5307–5318 (1986)
5.32 C. Koenig, E. Daniel: J. de Phys. **42**, L193–L195 (1981)
5.33 C. Koenig, P. Léonard, E. Daniel: J. de Phys. **42**, 1015–1023 (1981)
5.34 H.L. Skriver: *The LMTO Method*, Springer Ser. Solid-State Sci., Vol. 41 (Springer, Berlin, Heidelberg 1984)
5.35 O. Gunnarsson, O. Jepsen, O.K. Andersen: Phys. Rev. B **27**, 7144–7168 (1983)
5.36 A.V. Postnikov, V.P. Antropov, O. Jepsen: J. Phys.: Condensed Matter **4**, 2475–2486 (1992)
5.37 R. Zeller, J. Deutz, P.H. Dederichs: Solid State Commun. **44**, 993–997 (1982)
5.38 A.R. Mackintosh, O.K. Andersen: The electronic structure of transition metals, in *Electrons at the Fermi surface* (Cambridge University Press, Cambridge 1980) pp. 149–224
5.39 V. Heine: Electronic structure from the point of view of the local atomic environment, in *Solid State Physics*, Vol. 35, ed. by H. Ehrenreich, F. Seitz, P. Turnbull (Academic, New York 1980) pp. 1–128
5.40 M. Methfessel, J. Kübler: J. Phys. F **12**, 141–161 (1982)
5.41 P. Lloyd: Proc. Phys. Soc. London **90**, 207–216 (1967)
5.42 W. Heisenberg: Z. Phys. **49**, 619–636 (1928)
5.43 A.I. Liechtenstein, M.I. Katsnelson, V.A. Gubanov: Solid State Commun. **54**, 327–329 (1985)
5.44 A.I. Liechtenstein, M.I. Katsnelson, V.P. Antropov, V.A. Gubanov: J. Magn. Magn. Mater. **67**, 65–74 (1987)
5.45 J.S. Faulkner, G.M. Stocks: Phys. Rev. B **21**, 3222–3244 (1980)
5.46 A.J. Pindor, J. Staunton, G.M. Stocks, H. Winter: J. Phys. F **13**, 979–989 (1983)
5.47 A.I. Liechtenstein, M.I. Katsnelson, V.A. Gubanov: J. Phys. F **14**, L125–L128 (1984)
5.48 T. Oguchi, K. Terakura, N. Hamada: J. Phys. F **13**, 145–160 (1983)
5.49 B.L. Gyorffy, A.J. Pindor, J. Staunton, G.M. Stocks, H. Winter: J. Phys. F **15**, 1337–1386 (1985)
5.50 J. Staunton, B.L. Gyorffy, G.M. Stocks, J. Wadsworth: J. Phys. F **16**, 1761–1788 (1988)

5.51 R. Feder, F. Rosicky, B. Ackerman: Z. Phys. B **52**, 31–36 (1983)
5.52 H. Ebert, P. Strange, B.L. Gyorffy: J. Phys. F **18**, L135–L139 (1988)
5.53 H. Ebert: Phys. Rev. B **38**, 9390–9397 (1988)
5.54 B.C.H. Krutzen, F. Springelkamp: J. Phys.: Cond. Matter **1**, 8369–8383 (1989)
5.55 J. Banhart, P. Weinberger, J. Voitländer: J. Phys.: Cond. Matter. **1**, 7013–7020 (1989)
5.56 B. Ginatempo, J.B. Staunton: J. Phys. F **18**, 1827–1837 (1988)
5.57 G.M. Florio, B. Ginatempo, E.S. Giuliano, J.B. Staunton: J. Phys.: Cond. Matter **1**, 8385–8396 (1989)
5.58 T. Moriya: Phys. Rev. **120**, 91–98 (1960)

Chapter 6

6.1 H.J.F. Jansen, K.B. Hathaway, A.J. Freeman: Phys. Rev. B **30**, 6177–6179 (1979)
6.2 K.B. Hathaway, H.J.F. Jansen, A.J. Freeman: Phys. Rev. B **31**, 7603–7641
6.3 C.S. Wang, B.M. Klein, H. Krakauer: Phys. Rev. Lett. **54**, 1852–1855 (1985)
6.4 P. Bagno, O. Jepsen, O. Gunnarsson: Phys. Rev. B **40**, 1997–2000 (1989)
6.5 D.C. Langreth, M.J. Mehl: Phys. Rev. Lett. **47**, 446–449 (1981)
6.6 D.C. Langreth, M.J. Mehl: Phys. Rev. B **28**, 1809–1834 (1983)
6.7 C.D. Hu, D.C. Langreth: Phys. Rev. B **33**, 943–959 (1986)
6.8 J.P. Perdew: Phys. Rev. B **33**, 8822–8824 (1986)
6.9 J.P. Perdew: Phys. Rev. B **34**, 7407–7412 (1986)
6.10 J.P. Perdew, Y. Wang: Phys. Rev. B **33**, 8800–8802 (1986)
6.11 J. Kübler: Phys. Lett. A **81**, 81–83 (1981)
6.12 V.L. Moruzzi, J.F. Janak, A.R. Williams: *Calculated Electronic Properties of Metals* (Pergamon, Oxford 1978)
6.13 D.E. Eastman, F.J. Himpsel, J.A. Knapp: Phys. Rev. Lett. **40**, 1514–1517 (1978)
6.14 H. Eckardt, L. Fritsche: J. Phys. F **17**, 925–941 (1987)
6.15 B. Szpunar, P. Strange: J. Phys. F **15**, L165–L169 (1985)
6.16 V.P. Antropov: Electronic structure and magnetic interactions in transition metal alloys. Thesis, Sverdlovsk (1987)
6.17 V.L. Moruzzi, P.M. Marcus, K. Schwarz: Phys. Rev. B **33**, 1784–1791 (1986)
6.18 V.L. Moruzzi: Phys. Rev. Lett. **57**, 2211–2215 (1986)
6.19 K. Schwarz, P. Mohn: J. Phys. F **14**, L129–L134 (1984)
6.20 L.D. Landau, E.M. Lifshitz: *Statistical Physics* (Pergamon, New York 1980)
6.21 D. Bagayoko, J. Callaway: Phys. Rev. B **28**, 5419–5422 (1983)
6.22 S.C. Abrahams, L. Guttman, J.S. Kasper: Phys. Rev. B **127**, 2052–2055 (1962)
6.23 G. Johanson, M.B. McGirr, D.A. Wheeler: Phys. Rev. B **1**, 3208–3209 (1970)
6.24 P. Ehrhardt, B. Schonfeld, H.H. Ettwig, W. Pepperhoff: J. Magn. Magn. Mater. **22**, 79–85 (1980)
6.25 U. Gonser, K. Krischel, S. Nasu: J. Magn. Magn. Mater. **15–18**, 1145–1146 (1980)
6.26 A.J. Pindor, J. Staunton, G.H. Stocks, H. Winter: J. Phys. F **13**, 979–989 (1983)
6.27 B.L. Gyorffy, A.J. Pindor, J. Staunton, G.M. Stocks, H. Winter: J. Phys. F **15**, 1337–1386 (1985)
6.28 F.J. Pinski, J. Staunton, B.L. Gyorffy, D.D. Johnson, G.M. Stocks: Phys. Rev. Lett. **56**, 2096–2099 (1986)
6.29 J. Staunton, B.L. Gyorffy, G.H. Stocks, J. Wadsworth: J. Phys. F **16**, 1761–1788 (1986)
6.30 W. Williams, J.L. Stzanword: Phys. Rev. B **7**, 3244–3249 (1973)
6.31 N.I. Kulikov, V.V. Tugushev: Usp. Fiz. Nauk **144**, 643–680 (1984) [Engl. transl.: Sov. Phys. – Usp. **27**, 954 (1984)]
6.32 J. Kübler: J. Magn. Magn. Mater. **20**, 107–110 (1980)
6.33 J. Kübler: J. Magn. Magn. Mater. **20**, 277–284 (1980)
6.34 H.L. Skriver: J. Phys. F **11**, 97–113 (1981)
6.35 N.I. Kulikov, M. Alouani, M.A. Khan, M.V. Magnitskaya: Phys. Rev. B **36**, 924–938 (1987)
6.36 T. Jarlborg, A.J. Freeman: Phys. Rev. B **22**, 2332–2342 (1980)
6.37 A.R. Williams, V.L. Moruzzi, A.P. Malozemoff, K. Terakura: IEEE Trans. MAG-**19**, 1983–1988 (1983)
6.38 S. Chikazumi: *Physics of Magnetism* (Wiley, New York 1964)
6.39 A.P. Malozemoff, A.R. Williams, V.L. Moruzzi: Phys. Rev. B **29**, 1620–1632 (1984)

6.40 A.R. Williams, A.P. Malozemoff, V.L. Moruzzi, M. Matsui: J. Appl. Phys. **55**, 2353–2355 (1984)
6.41 J. Kübler: Physics B **127**, 257–263 (1984)
6.42 K. Schwarz, P. Mohn, P. Blaha, J. Kübler: J. Phys. F **14**, 2659–2671 (1984)
6.43 A.R. Williams, V.L. Moruzzi, C.D. Gelatt, Jr., J. Kübler: J. Magn.Magn. Mater. **31–34**, 88–94 (1983)
6.44 J. Kübler, A.R. Williams, C.B. Sommers: Phys. Rev. B **28**, 1745–1755 (1983)
6.45 A.R. Williams, R. Zeller, V.L. Moruzzi, C.D. Gelatt, Jr.: J. Appl. Phys. **52**, 2067–2069 (1981)
6.46 A.R. Williams, V.L. Moruzzi, C.D. Gelatt, Jr., J. Kübler, K. Schwarz: J. Appl. Phys. **53**, 2019–2023 (1982)
6.47 N.F. Mott: Proc. Phys. Soc. London A **62**, 416–422 (1949)
6.48 D. Adler: Insulating and metallic state in transition metal oxides, in *Solid State Physics*, Vol. 21, ed. by F. Seitz, D. Turnbull (Academic, New York 1968) p. 1
6.49 B.H. Brandow: Adv. Phys. **26**, 651–808 (1977)
6.50 A.B. Kunz, G.T. Surratt: Solid State Commun. **25**, 2989–2990 (1978)
6.51 A.B. Kunz: Int. J. Quantum Chem., Symp. **15**, 487 (1981)
6.52 B. Koiller and L.M. Falikov: J. Phys. C **7**, 289–295 (1974)
6.53 P.W. Anderson: Theory of magnetic exchange interactions: Exchange in insulators and semiconductors, in *Solid State Physics*, Vol. 14, ed. by F. Seitz, D. Turnbull (Academic, New York 1963) p. 99
6.54 B. Koiller, L.M. Falikov: J. Phys. C **8**, 695–707 (1975)
6.55 L.F. Mattheiss: Phys. Rev. B **5**, 290 (1972)
6.56 L.F. Mattheiss: Phys. Rev. B **5**, 306 (1972)
6.57 Z.-X. Shen, P.A.P. Lindberg, C.K. Shih, W.E. Spicer, I. Lindau: Physica C **162–164**, 1311–1312 (1989)
6.58 O.K. Andersen, H.L. Skriver, H. Nohl, B. Johansson: Pure Appl. Chem. **52**, 93 (1979)
6.59 T. Oguchi, K. Terakura, A.R. Williams: Phys. Rev. B **28**, 6443–6452 (1983)
6.60 T. Oguchi, K. Terakura, A.R. Williams: J. Appl. Phys. **55**, 2318–2320 (1984)
6.61 K. Terakura, T. Oguchi, A.R. Williams, J. Kübler: Phys. Rev. B **30**, 4734–4747 (1984)
6.62 J. Kübler, A.R. Williams: J. Magn. Magn. Mater. **54–57**, 603–606 (1986)
6.63 G.A. Sawatzky, J.W. Allen: Phys. Rev. Lett. **53**, 2339–2342 (1984)
6.64 J.M. McKay, V.E. Henrich: Phys. Rev. Lett. **53**, 2343–2345 (1984)
6.65 V.I. Anisimov, M.A. Korotin, I.V. Afanasyev: Physica C **161**, 59–65 (1989)
6.66 K. Yosida, A. Okiji, S. Chikazumi: Prog. Theor. Phys. **33**, 559–574 (1965)
6.67 J.P. Perdew, A. Zunger: Phys. Rev. B **23**, 5048–5079 (1981)
6.68 A. Svane, O. Gunnarsson: Europhys. Lett. **7**, 171–175 (1988)
6.69 J.S. Smart: Evaluation of exchange interactions from experimental data, in *Magnetism*, Vol. 3, ed. by G.T. Rado, H. Suhl (Academic, New York 1963) p. 63
6.70 M. Kohgi, Y. Ishikawa, Y. Endoh: Solid State Commun. **11**, 391 (1972)
6.71 M.E. Lines, E.D. Jones: Phys. Rev. B **139**, A313 (1965)
6.72 B.A. Coles, J.W. Orton, J. Owen: Phys. Rev. Lett. **4**, 116 (1960)
6.73 M.T. Hutchings, E.J. Samuelson: Solid State Commun. **9**, 1011–1014 (1971)
6.74 P.W. Andersen: Science **235**, 1196–1198 (1987)
6.75 G. Chen, W.A. Goddard: Science **239**, 899–902 (1988)
6.76 T. Freltoft, J.E. Fischer, G. Shirane, D.E. Moncton, S.K. Sinha, D. Vaknin, J.P. Remeika, A.S. Cooper, D. Harshman: Phys. Rev. B **36**, 826 (1987)
6.77 R.J. Birgeneau, G. Shirane: Neutron scattering studies of structural and magnetic excitations in lamellar copper oxides, *A Review of Physical Properties of High Temperature Superconductors*, ed. by D.M. Ginsberg (World Scientific, Singapore 1989) p. 151
6.78 Y. Endoh et al.: Phys. Rev. B **37**, 7443–7453 (1988)
6.79 G. Shirane: Proc. Int. Conf. Magnetism ICM-88, Paris, p. 296
6.80 J.M. Tranquada, D.E. Cox, W. Kunnmann, H. Moudden, G. Shirane, M. Suenaga, P. Zolliker: Phys. Rev. Lett. **60**, 156–159 (1988)
6.81 J.H. Brewer et al.: Phys. Rev. Lett. **60**, 1073–1076 (1988)
6.82 H. Kadowaki, M. Nishi, Y. Yamada, H. Takeya, H. Takei, S.M. Shapiro, BG. Shirane: Phys. Rev. B **37**, 7932 (1988)
6.83 T.C. Leung, X.W. Wang, B.N. Harmon: Phys. Rev. B **37**, 384–388 (1988)
6.84 K.A. Shiraishi, A. Oshigama, N. Shima, T. Nakayama, H. Kamimura: Solid State Commun. **69**, 629–632 (1988)
6.85 Y. Hatsugai, T. Fujiwara: Solid State Commun. **65**, 1271–1274 (1988)

6.86 G.Y. Guo, W.M. Temmerman, G.M. Stocks: J. Phys. C **21**, L103–L108 (1988)
6.87 W.E. Pickett, H. Krakauer, R.E. Cohen, D. Singh: Physica C **162–164**, 1419–1432 (1989)
6.88 D.L. Novikov, A.L. Ivanovsky, V.A. Gubanov: Fiz. Met. Metalloved. **64**, 195 (1987)
6.89 B. Szpunar, V.H. Smith, R.W. Smith: Physica C **152**, 91 (1988)
6.90 W.M. Temmerman, Z. Szotek, G.Y. Guo: J. Phys. C **21**, L807–L874 (1988)
6.91 A.K. McMahan, R.M. Martin, S. Satpathy: Phys. Rev. B **38**, 6650–6666 (1988)
6.92 M. Schluter, M.S. Hybertsen, N.E. Christensen: Physica C **153–155**, 1217 (1988)
6.93 M. Schluter, M.S. Hybertsen, N.E. Christensen: Phys. Rev. B **39**, 9029–9041 (1989)
6.94 J. Zaanen, T. Paxton, O. Jepsen, O.K. Andersen: Phys. Rev. Lett. **60**, 2685–2688 (1988)
6.95 H. Chen, J. Callaway, P.K. Misra: Phys. Rev. B **38**, 195–203 (1988)
6.96 Y. Guo, J.-M. Langlois, W.A. Goddard: Science **239**, 896–902 (1988)
6.97 V.A. Gubanov, S.A. Turzhevski, A.I. Liechtenstein, D.L. Novikov, V.I. Anisimov: Physica C **153–155**, 123 (1988)
6.98 A.R. Mackintosh, O.K. Andersen: The electronic structure of transition metals, in *Electrons at the Fermi Surface*, ed. by M. Springford (Cambridge University Press, Cambridge 1980) p. 149
6.99 O. Gunnarsson, O. Jepsen, O.K. Andersen: Phys. Rev. B **27**, 7144 (1988)
6.100 V.I. Anisimov, A.L. Ivanovsky, A.I. Liechtenstein, V.A. Gubanov: Zh. Strukt. Khim. **29**, 3–10 (1988)
6.101 B.A. Alekasashin et al.: Pis'ma Zh. Eksp. Teor. Fiz. **48**, 263 (1988) [Engl. transl.: Sov. Phys. – JETP Lett. **48**, 289–293 (1988)]
6.102 M. Weinert, G.W. Fernando: Phys. Rev. B **39**, 835–838 (1989)
6.103 T. Freltoft, G. Shirane, S. Mitsuda, J.P. Remeika, A.S. Cooper: Phys. Rev. B **37**, 137–142 (1988)

Chapter 7

7.1 V.L. Moruzzi, J.F. Janak, A.R. Williams: *Calculated Electronic Properties of Metals* (Pergamon, Oxford 1978)
7.2 J. Friedel: Nuovo Cimento, Suppl. Ser. X **7**, 287–311 (1958)
7.3 P.W. Anderson: Phys. Rev. **124**, 41–53 (1961)
7.4 P. Steiner, H. Höchst, S. Hüfner: J. Phys. F **7**, L105–L109 (1977)
7.5 P. Steiner, H. Höchst, W. Steffen, S. Hüfner: Z. Phys. B **38**, 191–199 (1980)
7.6 E. Mrosan, G. Lehmann: Phys. Status Solidi (b) **77**, K161–K164 (1976)
7.7 E. Mrosan, G. Lehmann: Phys. Status Solidi (b) **77**, 607–614 (1976)
7.8 E. Mrosan, G. Lehmann: Phys. Status Solidi (b) **87**, K21–K24 (1978)
7.9 R.M. Nieminen, M. Puska: J. Phys. F **10**, L123–L127 (1980)
7.10 G. Grüner: Adv. Phys. **23**, 941–1024 (1974)
7.11 R. Schöpke, E. Mrosan: Phys. Status Solidi (b) **90**, K95–K102 (1978)
7.12 J. Deutz, P.H. Dederichs, R. Zeller: J. Phys. F **11**, 1787–1800 (1981)
7.13 W.E. Pickett, B.M. Klein: Phys. Rev. B **29**, 1588–1596 (1984)
7.14 P. Léonard, N. Stefanou: Philos. Mag. B **51**, 151–160 (1985)
7.15 O. Gunnarsson: J. Phys. F **6**, 587–606 (1976)
7.16 R. Zeller, P.H. Dederichs: Phys. Rev. Lett. **42**, 1713–1716 (1979)
7.17 R. Zeller, R. Podloucky, P.H. Dederichs: Z. Phys. B **38**, 165–168 (1980)
7.18 R. Podloucky, R. Zeller, P.H. Dederichs: Phys. Rev. B **22**, 5777–5790 (1980)
7.19 R. Zeller, P.J. Braspenning: Solid State Commun. **42**, 701–704 (1982)
7.20 P.J. Braspenning, R. Zeller, A. Lodder, P.H. Dederichs: Phys. Rev. B **29**, 703–718 (1984)
7.21 J.C. Ododo: J. Phys. F **13**, 1291–1309 (1983)
7.22 B.H. Verbeck, G.N. Nieuwenhoys, J.A. Mydosh, C. van Dijk, B.D. Raiaford: Phys. Rev. B **22**, 5426–5440 (1980)
7.23 T.D. Cheung, J.S. Kouvel, J.W. Garland: Phys. Rev. B **23**, 1245–1252 (1981)
7.24 J.W. Loram, K.A. Mirza: J. Phyz. F **15**, 2213–2223 (1985)
7.25 B. Delley, D.E. Ellis: J. Magn. Magn. Mater. **30**, 71–86 (1982)
7.26 A. Oswald, R. Zeller, P.H. Dederichs: Phys. Rev. Lett. **56**, 1419–1422 (1986)
7.27 A. Oswald, R. Zeller, P.H. Dederichs: Magnetische 3d-Fremdatome und ihre Wechselwirkung in Kupfer. Silber und Palladium. Ber. Kernforschungsanlage Jülich N 2015 (1985)
7.28 J.W. Loram, K.A. Mirza: J. Phys. F **15**, 2213–2228 (1985)
7.29 T. Moriya: In *Theory of Magnetism in Transition Metals*, Proc. Int. School of Phys. Enrico Fermi, Vol. 37, ed. by W. Marshall (Academic, New York 1967) p. 206

7.30 T. Jarlborg, A.J. Freeman: Phys. Rev. B **23**, 3577–3579 (1981)
7.31 B.T. Mattheiss, M. Peter, H.J. Williams, A.M. Clogston, E. Corenzwit, R.C. Sherwood: Phys. Rev. Lett. **5**, 542–544 (1960)
7.32 A.M. Clogston, B.T. Mattheiss, M. Peter, H.J. Williams, E. Corenzwit, R.C. Sherwood: Phys. Rev. **125**, 541–552 (1962)
7.33 R. Podloucky, J. Deutz, R. Zeller, P.H. Dederichs: Phys. Status Solidi (b) **112**, 515–524 (1982)
7.34 J. Kübler: J. Magn. Magn. Mater. **20**, 107–110 (1980)
7.35 S. Ellialtioglu, R. Zeller, P.H. Dederichs: J. Phys. F **17**, 407–416 (1987)
7.36 S.V. Vonsovsky, M.I. Katsnelson, A.V. Trefilov: J. Magn. Magn. Mater. **61**, 83–87 (1986)
7.37 J.A. Cape, R.R. Hake: Phys. Rev. **139**, A142–A149 (1965)
7.38 A.S. Shcherbakov, N.V. Volkenshtein, E.G. Valiulin, N.Ya. Permyakov: Pis'ma Zh. Eksp. Teor. Fiz. **36**, 320–322 (1982) [Engl. transl.: Sov. Phys. – JETP Lett. **36**, 390–393 (1982)]
7.39 A.S. Sherbakov, M.I. Katsnelson, A.V. Trefilov, E.N. Bulatov, N.V. Volkenshtein, E.G. Valiulin: Pis'ma Zh. Eksp. Ther. Fiz. **44**, 393–396 (1986) [Engl. transl.: Sov. Phys. – JETP Lett. **44**, 504–508 (1986)]
7.40 A.V. Postnikov, V.I. Anisimov, V.A. Gubanov: J. Magn. Magn. Mater. **39**, 295–308 (1983)
7.41 V.P. Antropov, V.I. Anisimov, M.I. Katsnelson, A.I. Liechtenstein, A.V. Trefilov: Fiz. Met. Metalloved. **68**, 845–853 (1989)
7.42 V.P. Antropov: Electronic structure and magnetic interactions in transition metal alloys. Thesis, Sverdlovsk (1987)
7.43 N. Stefanou, P.J. Braspenning, R. Zeller, P.H. Dederichs: Phys. Rev. B **36**, 6372–6382 (1987)
7.44 C. Koenig, E. Daniel: J. de Phys. **42**, 1015–1023 (1981)
7.45 P.H. Dederichs, R. Zeller: Electronic structure of impurities in transition metals, in *Advances in Solid Physics*, Vol. 21 (Vieweg, Braunschweig 1981) pp. 243–270
7.46 N. Stefanou, A. Oswald, R. Zeller, P.H. Dederichs: Phys. Rev. B **35**, 6911–6922 (1987)
7.47 M. Akai, H. Akai, J. Kanamori: J. Phys. Soc. Jpn. **54**, 4230–4245 (1985)
7.48 H. Akai, M. Akai, J. Kanamori: J. Phys. Soc. Jpn. **54**, 4246–4256 (1985)
7.49 V.I. Anisimov, V.P. Antropov, A.I. Liechtenstein, V.A. Gubanov, A.V. Postnikov: Phys. Rev. B **37**, 5598–5602 (1988)
7.50 H. Akai, M. Akai, S. Blügel, R. Zeller, P.H. Dederichs: J. Magn. Magn. Mater. **45**, 291–297 (1984)
7.51 S.G. Kaug, H. Onodera: J. Phys. Soc. Jpn. **36**, 978–981 (1974)
7.52 D. Bagayoko, J. Callaway: Phys. Rev. B **28**, 5419–5422 (1983)
7.53 D. Pescia, M. Stampanoni, G.L. Bona, A. Vateriaus, R.F. Willis, F. Meier: Phys. Rev. Lett. **58**, 2126–2129 (1987)
7.54 C. Sadron: Ann. de Phys. **17**, 371–452 (1932)
7.55 V. Marian: Ann. de Phys. **7**, 459–527 (1937)
7.56 N.M. March, Ph. Lambin, F. Herman: J. Magn. Magn. Mater. **44**, 1–19 (1984)
7.57 B. Lengeler: *Microstructural Characterization of Materials by Non-microscopical techniques*, ed. by N.H. Andersen (Risø National Lab., Roskilde 1984) pp. 363–370
7.58 R. Zeller: J. Phys. F **15**, 731–742 (1987)
7.59 V.R. Galakhov, E.Z. Kurmaev, A.V. Postnikov: Solid State Commun. **58**, 143–146 (1986)
7.60 V.A. Gubanov, A.I. Liechtenstein, A.V. Postnikov: Int. J. Quant. Chem. **23**, 1517–1528 (1983)
7.61 Y. Ito, J. Akimitsu: J. Phys. Soc. Jpn. **36**, 431–437, 1974
7.62 L. Madhav, Rao, R. Chakravartym: Z. Jirak, N.S. Satya Murthy: Phys. Rev. B **18**, 6275–6282 (1978)
7.63 I.A. Campell, A.L. Allsop, N.I. Stone: J. Phys. F **8**, L235–L237 (1978)
7.64 J.W. Cable: Phys. Rev. B **25**, 4670–4679 (1982)
7.65 T. Miyamoto, Y. Nakai, N. Kanitomu: J. Phys. Soc. Jpn. **54**, 4367–4372 (1984)
7.66 N.I. Kulikov, V.V. Tugushev: Usp. Fiz. Nauk **144**, 643–680 (1984) [Engl. transl.: Sov. Phys. – Usp. **27**, 954–976 (1984)]
7.67 W.M. Lomer: Proc. Phys. Soc. **80**, 489–497 (1962)
7.68 Y. Ishikawa, R. Tournier, J. Flippi: J. Phys. Chem. Solids **26**, 1727–1733 (1965)
7.69 W. Williams, J.L. Stanword: Phys. Rev. B **7**, 3244–3249 (1973)
7.70 F.J. Hedgcock, J.O. Strom-Olsen, D.F. Wilford: J. Phys. F **7**, 855–860 (1977)
7.71 J. Kübler: J. Magn. Magn. Mater. **20**, 277–284 (1980)
7.72 N.I. Kulikov, M. Alouani, M.A. Khan, M.V. Magnitskaya: Phys. Rev. B **36**, 924–938 (1987)
7.73 D.R. Salahub, R.P. Messmer: Surf. Sci. **106**, 415–421 (1981)
7.74 T.M. Rice, A.S. Barker, Jr., B.I. Halperin, D.B. McWhan: J. Appl. Phys. **40**, 1337–1343 (1969)
7.75 J. Kübler: J. Magn. Magn. Mater. **20**, 107–110 (1980)

7.76 V.I. Anisimov, V.P. Antropov, V.A. Gubanov, A.I. Ivanovsky, E.Z. Kurmaev, A.I. Liechtenstein, A.V. Postnikov: *Electronic Structure of Impurities in Transition Metals and Their Compounds* (Nauka, Moscow 1989) (In Russian)
7.77 V.P. Antropov, V.I. Anisimov, A.I. Liechtenstein, A.V. Postnikov: Phys. Rev. B **37**, 5603–5605 (1988)
7.78 A.V. Postnikov, V.I. Anisimov: Fiz. Met. Metalloved. **67**, 845–853 (1989)
7.79 J. Kaspar, D.R. Salahub: J. Phys. F **13**, 311–332 (1983)
7.80 K. Schwarz, P. Mohn, P. Blaha, J. Kübler: J. Phys. F **14**, 2659–2671 (1984)
7.81 C.M. Hord: J. Phys. Chem. Solids **28**, 1345–1352 (1967)
7.82 P.J. Ford: Contemp. Phys. **23**, 141–168 (1982)
7.83 A. Oswald, R. Zeller, P.J. Braspenning, P.H. Dederichs: J. Phys. F **15**, 193–212 (1985)
7.84 P.H. Dederichs, R. Zeller, H. Akai, S. Blügel, A. Oswald: Philos. Mag. B **51**, 137–150 (1985)
7.85 J.R. Davis, S.K. Burke, B.D. Rainford: J. Phys. F **9**, L7–L10 (1979)
7.86 J.R. Davis, T.J. Hichs: J. Phys. F **9**, 753–761 (1979)
7.87 J.R. Davis, S.K. Burke, B.D. Rainford: J. Magn. Magn. Mater. **15–18**, 151–152 (1980)
7.88 J.C. Parlebas, F. Gautier: J. Phys. F **4**, 1044–1064 (1974)
7.89 A.F.J. Morgownik, J.A. Mydosh: Solid State Commun. **47**, 321–324 (1983)
7.90 A. Oswald, R. Zeller, P.H. Dederichs: J. Magn. Magn. Mater. **54–57**, 1247–1248 (1986)

Subject Index

Index of Compounds

Springer Series in Solid-State Sciences

Editors: M. Cardona · P. Fulde · K. von Klitzing · H.-J. Queisser

1 Principles of Magnetic Resonance
3rd Edition By C. P. Slichter

2 Introduction to Solid-State Theory
2nd Printing By O. Madelung

3 Dynamical Scattering of X-Rays in Crystals By Z. G. Pinsker

4 Inelastic Electron Tunneling Spectroscopy
Editor: T. Wolfram

5 Fundamentals of Crystal Growth I
Macroscopic Equilibrium and Transport Concepts. 2nd Printing
By F. E. Rosenberger

6 Magnetic Flux Structures in Superconductors By R. P. Huebener

7 Green's Functions in Quantum Physics
2nd Edition By E. N. Economou

8 Solitons and Condensed Matter Physics
2nd Printing
Editors: A. R. Bishop and T. Schneider

9 Photoferroelectrics By V. M. Fridkin

10 Phonon Dispersion Relations in Insulators By H. Bilz and W. Kress

11 Electron Transport in Compound Semiconductors By B. R. Nag

12 The Physics of Elementary Excitations
By S. Nakajima, Y. Toyozawa, and R. Abe

13 The Physics of Selenium and Tellurium
Editors: E. Gerlach and P. Grosse

14 Magnetic Bubble Technology 2nd Edition
By A. H. Eschenfelder

15 Modern Crystallography I
Symmetry of Crystals.
Methods of Structural Crystallography
By B. K. Vainshtein

16 Organic Molecular Crystals
Their Electronic States By E. A. Silinsh

17 The Theory of Magnetism I
Statics and Dynamics. 2nd Printing
By D. C. Mattis

18 Relaxation of Elementary Excitations
Editors: R. Kubo and E. Hanamura

19 Solitons, Mathematical Methods for Physicists. 2nd Printing
By G. Eilenberger

20 Theory of Nonlinear Lattices
2nd Edition By M. Toda

21 Modern Crystallography II
Structure of Crystals
By B. K. Vainshtein, V. M. Fridkin, and V. L. Indenbom

22 Point Defects in Semiconductors I
Theoretical Aspects
By M. Lannoo and J. Bourgoin

23 Physics in One Dimension
Editors: J. Bernasconi, T. Schneider

24 Physics in High Magnetic Fields
Editors: S. Chikazumi and N. Miura

25 Fundamental Physics of Amorphous Semiconductors Editor: F. Yonezawa

26 Elastic Media with Microstructure I
One-Dimensional Models. By I. A. Kunin

27 Superconductivity of Transition Metals
Their Alloys and Compounds
By S. V. Vonsovsky, Yu. A. Izyumov, and E. Z. Kurmaev

28 The Structure and Properties of Matter
Editor: T. Matsubara

29 Electron Correlation and Magnetism in Narrow-Band Systems Editor: T. Moriya

30 Statistical Physics I
Equilibrium Statistical Mechanics
By M. Toda, R. Kubo, N. Saito

31 Statistical Physics II
Nonequilibrium Statistical Mechanics
By R. Kubo, M. Toda, N. Hashitsume

32 Quantum Theory of Magnetism
2nd Edition By R. M. White

33 Mixed Crystals By A. I. Kitaigorodsky

34 Phonons: Theory and Experiments I
Lattice Dynamics and Models of Interatomic Forces By P. Brüesch

35 Point Defects in Semiconductors II
Experimental Aspects
By J. Bourgoin and M. Lannoo

36 Modern Crystallography III
Crystal Growth. 2nd Edition
By A. A. Chernov

37 Modern Crystallography IV
Physical Properties of Crystals
Editor: L. A. Shuvalov

38 Physics of Intercalation Compounds
Editors: L. Pietronero and E. Tosatti

39 Anderson Localization
Editors: Y. Nagaoka and H. Fukuyama

40 Semiconductor Physics An Introduction
3rd Edition By K. Seeger

41 The LMTO Method
Muffin-Tin Orbitals and Electronic Structure
By H. L. Skriver

42 Crystal Optics with Spatial Dispersion, and Excitons 2nd Edition
By V. M. Agranovich and V. L. Ginzburg

43 Structure Analysis of Point Defects in Solids
An Introduction to Multiple Magnetic Resonance Spectroscopy
By J.-M. Spaeth, J. R. Niklas, and R. H. Bartram

44 Elastic Media with Microstructure II
Three-Dimensional Models By I. A. Kunin

45 Electronic Properties of Doped Semiconductors
By B. I. Shklovskii and A. L. Efros

46 Topological Disorder in Condensed Matter
Editors: F. Yonezawa and T. Ninomiya

47 Statics and Dynamics of Nonlinear Systems
Editors: G. Benedek, H. Bilz, and R. Zeyher

48 Magnetic Phase Transitions
Editors: M. Ausloos and R. J. Elliott

49 Organic Molecular Aggregates Electronic Excitation and Interaction Processes
Editors: P. Reineker, H. Haken, and H. C. Wolf

Springer Series in Solid-State Sciences

Editors: M. Cardona P. Fulde K. von Klitzing H.-J. Queisser

1 **Principles of Magnetic Resonance** 3rd Edition By C. P. Slichter

2 **Introduction to Solid-State Theory** 2nd Printing By O. Madelung

3 **Dynamical Scattering of X-Rays in Crystals** By Z. G. Pinsker

4 **Inelastic Electron Tunneling Spectroscopy** Editor: T. Wolfram

5 **Fundamentals of Crystal Growth I** Macroscopic Equilibrium and Transport Concepts. 2nd Printing By F. Rosenberger

6 **Magnetic Flux Structures in Superconductors** By R. P. Huebener

7 **Green's Functions in Quantum Physics** 2nd Edition By E. N. Economou

8 **Solitons and Condensed Matter Physics** 2nd Printing Editors: A. R. Bishop and T. Schneider

9 **Photoferroelectrics** By V. M. Fridkin

10 **Phonon Dispersion Relations in Insulators** By H. Bilz and W. Kress

11 **Electron Transport in Compound Semiconductors** By B. R. Nag

12 **The Physics of Elementary Excitations** By S. Nakajima, Y. Toyozawa, and R. Abe

13 **The Physics of Selenium and Tellurium** Editors: E. Gerlach and P. Grosse

14 **Magnetic Bubble Technology** 2nd Edition By A. H. Eschenfelder

15 **Modern Crystallography I** Symmetry of Crystals. Methods of Structural Crystallography By B. K. Vainshtein

16 **Organic Molecular Crystals** Their Electronic States. By E. A. Silinsh

17 **The Theory of Magnetism I** Statics and Dynamics. 2nd Printing By D. C. Mattis

18 **Relaxation of Elementary Excitations** Editors: R. Kubo and E. Hanamura

19 **Solitons.** Mathematical Methods for Physicists. 2nd Printing By G. Eilenberger

20 **Theory of Nonlinear Lattices** 2nd Edition By M. Toda

21 **Modern Crystallography II** Structure of Crystals By B. K. Vainshtein, V. M. Fridkin, and V. L. Indenbom

22 **Point Defects in Semiconductors I** Theoretical Aspects By M. Lannoo and J. Bourgoin

23 **Physics in One Dimension** Editors: J. Bernasconi, T. Schneider

24 **Physics in High Magnetic Fields** Editors: S. Chikazumi and N. Miura

25 **Fundamental Physics of Amorphous Semiconductors** Editor: F. Yonezawa

26 **Elastic Media with Microstructure I** One-Dimensional Models. By I. A. Kunin

27 **Superconductivity of Transition Metals** Their Alloys and Compounds By S. V. Vonsovsky, Yu. A. Izyumov, and E. Z. Kurmaev

28 **The Structure and Properties of Matter** Editor: T. Matsubara

29 **Electron Correlation and Magnetism in Narrow-Band Systems** Editor: T. Moriya

30 **Statistical Physics I** Equilibrium Statistical Mechanics By M. Toda, R. Kubo, N. Saito

31 **Statistical Physics II** Nonequilibrium Statistical Mechanics By R. Kubo, M. Toda, N. Hashitsume

32 **Quantum Theory of Magnetism** 2nd Edition By R. M. White

33 **Mixed Crystals** By A. I. Kitaigorodsky

34 **Phonons: Theory and Experiments I** Lattice Dynamics and Models of Interatomic Forces. By P. Brüesch

35 **Point Defects in Semiconductors II** Experimental Aspects By J. Bourgoin and M. Lannoo

36 **Modern Crystallography III** Crystal Growth 2nd Edition By A. A. Chernov

37 **Modern Crystallography IV** Physical Properties of Crystals Editor: L. A. Shuvalov

38 **Physics of Intercalation Compounds** Editors: L. Pietronero and E. Tosatti

39 **Anderson Localization** Editors: Y. Nagaoka and H. Fukuyama

40 **Semiconductor Physics** An Introduction 5th Edition By K. Seeger

41 **The LMTO Method** Muffin-Tin Orbitals and Electronic Structure By H. L. Skriver

42 **Crystal Optics with Spatial Dispersion, and Excitons** 2nd Edition By V. M. Agranovich and V. L. Ginzburg

43 **Structure Analysis of Point Defects in Solids by Multiple Magnetic Resonance Spectroscopy** By J.-M. Spaeth, J. R. Niklas, and R. H. Bartram

44 **Elastic Media with Microstructure II** Three-Dimensional Models By I. A. Kunin

45 **Electronic Properties of Doped Semiconductors** By B. I. Shklovskii and A. L. Efros

46 **Topological Disorder in Condensed Matter** Editors: F. Yonezawa and T. Ninomiya

47 **Statics and Dynamics of Nonlinear Systems** Editors: G. Benedek, H. Bilz, and R. Zeyher

48 **Magnetic Phase Transitions** Editors: M. Ausloos and R. J. Elliott

49 **Organic Molecular Aggregates** Electronic Excitation and Interaction Processes Editors: P. Reineker, H. Haken, and H. C. Wolf

Springer Series in Solid-State Sciences

Editors: M. Cardona P. Fulde K. von Klitzing H.-J. Queisser

50 **Multiple Diffraction of X-Rays in Crystals**
By Shih-Lin Chang

51 **Phonon Scattering in Condensed Matter**
Editors: W. Eisenmenger, K. Laßmann, and S. Döttinger

52 **Superconductivity in Magnetic and Exotic Materials** Editors: T. Matsubara and A. Kotani

53 **Two-Dimensional Systems, Heterostructures, and Superlattices**
Editors: G. Bauer, F. Kuchar, and H. Heinrich

54 **Magnetic Excitations and Fluctuations**
Editors: S. Lovesey, U. Balucani, F. Borsa, and V. Tognetti

55 **The Theory of Magnetism II** Thermodynamics and Statistical Mechanics By D. C. Mattis

56 **Spin Fluctuations in Itinerant Electron Magnetism** By T. Moriya

57 **Polycrystalline Semiconductors,**
Physical Properties and Applications
Editor: G. Harbeke

58 **The Recursion Method and Its Applications**
Editors: D. Pettifor and D. Weaire

59 **Dynamical Processes and Ordering on Solid Surfaces** Editors: A. Yoshimori and M. Tsukada

60 **Excitonic Processes in Solids**
By M. Ueta, H. Kanzaki, K. Kobayashi, Y. Toyozawa, and E. Hanamura

61 **Localization, Interaction, and Transport Phenomena** Editors: B. Kramer, G. Bergmann, and Y. Bruynseraede

62 **Theory of Heavy Fermions and Valence Fluctuations** Editors: T. Kasuya and T. Saso

63 **Electronic Properties of Polymers and Related Compounds**
Editors: H. Kuzmany, M. Mehring, and S. Roth

64 **Symmetries in Physics** Group Theory Applied to Physical Problems
By W. Ludwig and C. Falter

65 **Phonons: Theory and Experiments II**
Experiments and Interpretation of Experimental Results By P. Brüesch

66 **Phonons: Theory and Experiments III**
Phenomena Related to Phonons
By I. Brüesch

67 **Two-Dimensional Systems: Physics and New Devices**
Editors: G. Bauer, F. Kuchar, and H. Heinrich

68 **Phonon Scattering in Condensed Matter V**
Editors: A. C. Anderson and J. P. Wolfe

69 **Nonlinearity in Condensed Matter**
Editors: A. R. Bishop, D. K. Campbell, P. Kumar, and S. E. Trullinger

70 **From Hamiltonians to Phase Diagrams** The Electronic and Statistical-Mechanical Theory of sp-Bonded Metals and Alloys By J. Hafner

71 **High Magnetic Fields in Semiconductor Physics**
Editor: G. Landwehr

72 **One-Dimensional Conductors**
By S. Kagoshima, H. Nagasawa, and T. Sambongi

73 **Quantum Solid-State Physics**
Editors: S. V. Vonsovsky and M. I. Katsnelson

74 **Quantum Monte Carlo Methods** in Equilibrium and Nonequilibrium Systems Editor: M. Suzuki

75 **Electronic Structure and Optical Properties of Semiconductors** 2nd Edition
By M. L. Cohen and J. R. Chelikowsky

76 **Electronic Properties of Conjugated Polymers**
Editors: H. Kuzmany, M. Mehring, and S. Roth

77 **Fermi Surface Effects**
Editors: J. Kondo and A. Yoshimori

78 **Group Theory and Its Applications in Physics**
By T. Inui, Y. Tanabe, and Y. Onodera

79 **Elementary Excitations in Quantum Fluids**
Editors: K. Ohbayashi and M. Watabe

80 **Monte Carlo Simulation in Statistical Physics**
An Introduction
By K. Binder and D. W. Heermann

81 **Core-Level Spectroscopy in Condensed Systems**
Editors: J. Kanamori and A. Kotani

82 **Introduction to Photoemission Spectroscopy**
By S. Hüfner

83 **Physics and Technology of Submicron Structures**
Editors: H. Heinrich, G. Bauer, and F. Kuchar

84 **Beyond the Crystalline State** An Emerging Perspective By G. Venkataraman, D. Sahoo, and V. Balakrishnan

85 **The Fractional Quantum Hall Effect**
Properties of an Incompressible Quantum Fluid
By T. Chakraborty and P. Pietilainen

86 **The Quantum Statistics of Dynamic Processes**
By E. Fick and G. Sauermann

87 **High Magnetic Fields in Semiconductor Physics II**
Transport and Optics Editor: G. Landwehr

88 **Organic Superconductors**
By T. Ishiguro and K. Yamaji

89 **Strong Correlation and Superconductivity**
Editors: H. Fukuyama, S. Maekawa, and A. P. Malozemoff

MIX
Papier aus verantwortungsvollen Quellen
Paper from responsible sources
FSC® C105338

If you have any concerns about our products,
you can contact us on
ProductSafety@springernature.com

In case Publisher is established outside the EU,
the EU authorized representative is:
Springer Nature Customer Service Center GmbH
Europaplatz 3, 69115 Heidelberg, Germany

Printed by Libri Plureos GmbH
in Hamburg, Germany

SpringerBriefs in Economics

For further volumes:
http://www.springer.com/series/8876